THE ART OF NUMBER

BY JAIN 108

Translating Numbers Into Pictures

and Sacred Symbol

2012

contact: jain@jainmathemagics.com

website: www.jainmathemagics.com

www.jain108.com

ISBN

978-1-925834-25-3

(The Front Cover Design "**Prime Number Cross**" is sourced from Page 157).

THE ART OF NUMBER
BY
JAIN 108

DEDICATION

"...to the pursuit and attainment of True Unconditional Love, beyond the acquisition of Knowledge..."

Jain 108

(Art By Jain, 2012. "Angelic Winged Lovers")

CONTENTS IN BRIEF

CONTENTS IN FULL

Art of Number

· The Translation of Number in2 Pictures.

· The Mathematical Derivation of Sacred Symbols.

~ P R E F A C E ~

+

~ INTRODUCTION ~
DIGITAL COMPRESSION
& THE REVELATION OF THE 108 PHI CODES

One of the Laws Of Nature is the ability to survive after **Compression**. This is why when we observe the counter rotating field of the sunflower floret, where there are 21 spirals going clockwise and 34 spirals going anti-clockwise ie: 21:34, the seeds do not fall out of their arrangement due to their intelligent, sophisticated and compact arrangement. Nature abhors midpoints and equality, you will never see a sunflower having an equal spin arrangement of 21:21.

In fact Nature prefers a frequency that is a fraction more than half (1/2) or Point 5 (.5) of Unity (1) known as the Phi Ratio or .618:1. In the Human Anatomy, it is the place where the elbow bends. The elbow does not bend at .5 or midpoint, it bends at a place that is a fraction more, called .618... (or approximately 21 divided by 34), its origins being embedded in the Vortex Mathematics or **Living Curvation** of the Sunflower and Pine Cone.

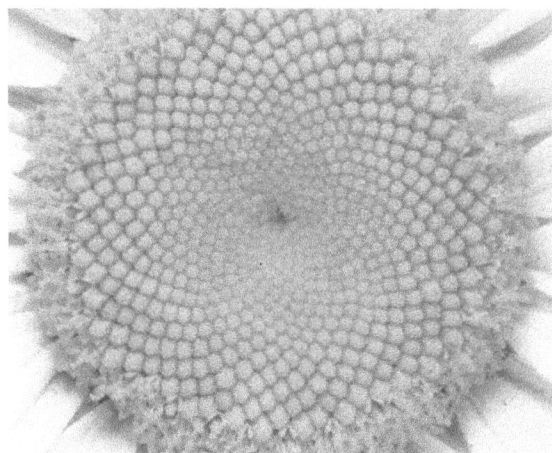

Fig 1
(Dimension 1)
Sunflower floret and face of Daisy, having 21:34 and 34:55 spirals
going clockwise and anticlockwise respectively.

These often reoccurring Numbers of Nature, are part of the Fibonacci Sequence: **1, 1, 2, 3, 5, 8, 13, 21, 34, 55, 89, 144** etc.

This sequence can be seen as a series of squares ever increasing in size. When an arc is passed through them, a beautiful spiral is formed

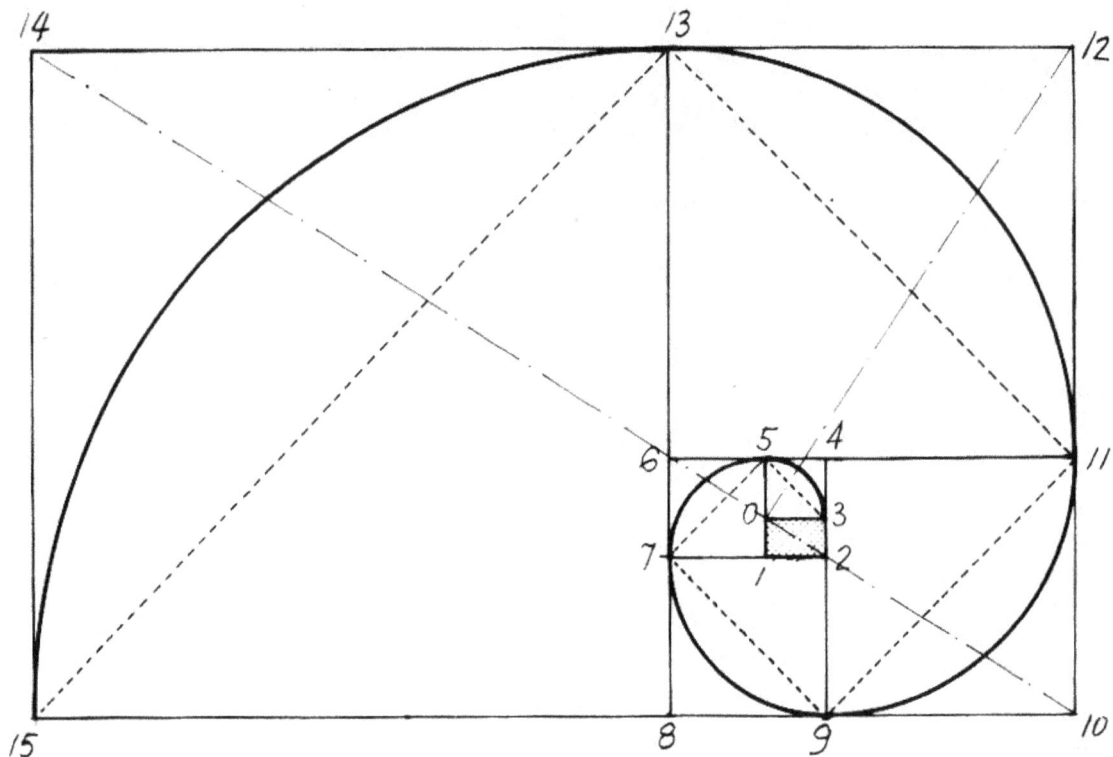

Fig 2

(Dimension 2)
This typical shell growth is known as a Logarithmic Spiral. Its successive stages of growth are indicated by the "whirling squares" and phi-ratioed golden rectangles that continually grow outwardly in harmonic progression from the "0" centre.

When number is turned into art, it is often visible as Atomic Art or in this case, The Living Mathematics of Nature. Shown is the 2-dimensional Golden Spiral, the familiar Nautilus Shell shape.

Fig 3

(Dimension 3)

When the flatland Squares are subsequently seen as Cubes,
and Arcs in Space are made progressively according to these Fibonacci
Numbers, that which results is the 3-dimensional analogue, recognized in
seashells or in this case, the exquisite curves of the Bighorn Sheep's horns.

When a student therefore understands symbolically the 1-Dimension of say the Sequence of the Fibonacci Numbers that generated from Squares and Arcs the 2-Dimensional Nautilus Spiral, they see how the Squares can become Cubes, and these Arcs in Cubes generated the Ram's Horn spiral, but the question is: **What is the 4th Dimensional analogue of this Stairway To Heaven?** The 4th Dimensional view of the Golden Spiral is a biological shape that we were in our 512 Cell Division of the zygote, called a Torus (the Latin word for "Ring"):

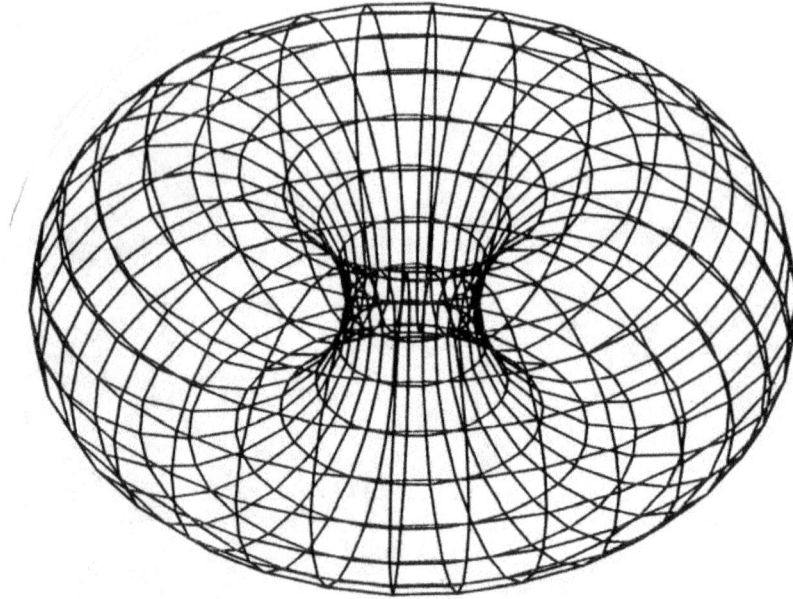

Fig 4
(Dimension 4)
The shift from 3 to 4 dimensions is the arrangement of an
infinite array of 2-dim golden spirals that generate the Torus,
the underlying shape of all creation from atoms to stars

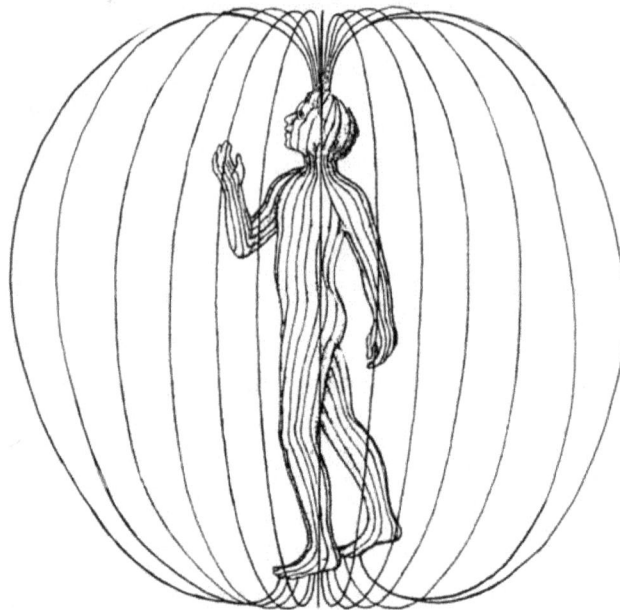

Fig 5
(Dimension 5)
Ultimately, in Sacred Geometry, it is not about having this Knowledge, it is about
embedding this Wisdom in our Hearts, Becoming and Living It.

This chapter on the Divine Proportion will be discussed in full, but first let us look
at some other simple examples.

The purpose of this book and the series to follow is to translate such number sequences (male left brain) and convert them into shapes or patterns (right brain mathematics) so that the student can perceive this knowledge from the perspective of Whole Brain Learning. In my other books, I gave this concept a new name, called "In The Next Dimension" or "Jain's 17th Sutra" a clever and much needed continuation of the 16 Sutras or Laws of Numbers expressed in Vedic Mathematics (Rapid Mental Calculation). We therefore use Base 10 in our daily world to efficiently calculate and build and solve problems, but it is not the only base of numbers, there are 3 more important ones discussed in this book, known as Base 2 (computers), Base 9 and Base 12 (Galactic Number Bases).

In the technological world, our computer language based on "0 and 1" is a form of Base 2 with On and Off switches. It is a form of Digital Compression.

Some numbers like 108 in our Base 10 system, which efficiently uses only 3 numbers in Base 10, involves 7 numbers in Base 2. We use Base 10 (that involves the numbers from 0, 1, 2, 3, 4, 5, 6, 7, 8, 9) as it is highly efficient compared to the important though long-winded Base 2 (that involves the numbers from 0, 1). Really, **our Decimal System is a Base 9 Modulus System!**

Here is an example shown in the chart for "108" in Base Ten and converted to Base 2 which is based on the successive Powers of 2 and represented as "1101100".

108							
2^7	2^6	2^5	2^4	2^3	2^2	2^1	2^0
128	64	32	16	8	4	2	1
	✓	✓		✓	✓		
	1	1	0	1	1	0	0

Fig 6

Our numbers in Base 10, like the example above, the number 108, uses only 3 digits to express itself, whereas when converted into Base 2, it uses 7 digits and looks like: 1101100. Mathematically, it is written as:

$108_{10} = 1101100_2$ where the subscripts indicate the bases used.

You can see that larger numbers in Base 10 like 432 will require more and more powers of two and therefore more numbers, so after thousands of years of trial and error, the world has adopted Base 10 as the most economical form of communication as in Rapid Mental Calculation aka Vedic Mathematics, but we still need to know Base 9 and Base 10 as that is where all the secrets lie. In the course of this book you will see how Nature chooses the Number 9 and the Number 12 to embed or hide the invisible, galactic knowledge. It is through this concept of Turning Numbers Into Shapes that makes the Invisible to appear as Visible. This alchemical shift happens in the crucible of our brain. Think of Numbers as abstracts, they live in the Left Hand Cerebral Portion of the Brain that governs analysis, rational, male logic, intellect. Understand that the human brain also works with a Language of Shape and Symbol that lives in the Right Feminine Cortex of the Brain, and governs Intuition, Holograms, Patterns, Creativity, Music etc.

So when your child is Joining The Dots of the Number Sequences given in this book, they are accessing Higher Dimensional Knowledge, they are working from a space called Whole Brain Learning. Literally they become switched on. Thus we need to teach our children this Artistic Maths first, then their brains will be funneled open to receive all the harder maths like algebra and calculus if they choose so at a later date. So we have been doing it back the front for all these centuries, our good-meaning and knowledgeable mathematics teachers were instructed to give us a watered-down curriculum, that needs urgently to be injected with The Art Of Numbers. This will create a race of geniuses.

The subtle reason why the **Number 108** was used as the previous example, was not arbitrary. It is the solution to a hidden and highly intelligent pattern that runs through the Fibonacci Sequence.
For over 2,000 years, and still today, we have been told that there is no Pattern in the Divine Proportion, yet this is not true. In summary, there exists an infinitely repeating 24 Pattern in this Fibonacci Sequence, a veritable Wheel of 24, a Time Code, a Stargate perhaps offering clues to the ancient Art of Time Bending, or the physics of Black Holes sufficiently realized by Turning Number Into Art. The missing step though is at the level of Digital Compression. To get to the cryptic pattern, the student must learn how to reduce larger numbers to single digits from 1 to 9, which is shown in the next paragraph heading.

Rather than showing you The Digitally Compressed Wheel of 24 as a linear 1 row of 24 digits, here it is in its 2x12 format:

1	1	2	3	5	8	4	3	7	1	8	9
8	8	7	6	4	1	5	6	2	8	1	9

Fig 7

The 108 Phi Code, 1 of 3 possible sequences (rediscovered by Jain 108 and other Mathematical Monks, Numerical Nomads and engrailed Phiometrists).

The Phi Code 108 is shown above in its 2x12 format, to illustrate how the 12 vertical columns or Pairs all have a Sum of 9. Basically, the sum of all the 12 Pairs of 9 is 12x9=**108**.

As mentioned earlier in this article, Base 9 and Base 12 in mathematics give us X-Ray eyes to understand the Fabric of Creation and Sacred Bio-Architecture of the Pyramids of Egypt and Stonehenge in the UK based on 12 inches in the Imperial Foot whose template of temple building was based on 12x12=144 inches to the square foot, **144** being an Anointed Number relating to the Harmonic of Light.

From this basic premise and revelation, the keen student will learn advanced Masonic or **Theo**-sophical ("Theo" = God, in Greek) Knowledge, understanding for example, how the sacred number 108 is **geometrically** associated to the number 144 as a sub-harmonic. The clue is based on a simple Universal Law of Numbers that states that **3x3 + 4x4 = 5x5**.

WHAT IS DIGITAL COMPRESSION?

The Digital Compression of Big Numbers is the simplification of its many digits a Single Digit from 1 to 9. Other names for **Digital Compression** include:

− Numerical Reduction

− Digital Reduction

− Reduced Numbers

− Casting Out Of Nines

− Distillation

− Digit Sums

A two digit number **13** like can be reduced to a single digit by either adding the separate digits as in 1+3=4 or by subtracting the number 9 from 13 which = 4. That's what Casting Out Of Nines really means. Just keep subtracting 9s from the number until a single digit is reached. This was pioneered by the great German mathematician Frederich Gauss in his modular arithmetic.

Really, we are saying that the vibration of "4" holds the memory of "13" in the same way that the outer or valence electrons around a complex atomic core determines the structure of the specific atomic element.

For bigger numbers like 4096, this will take too long, so best to add the digits: eg: **4096** = 4+0+9+6 =19 and add again 19=1+9=10 and add again 10=1+0=1. Therefore 4096 digitally compressed =1.

When a large number has lots of 9s in it, they can also be cast out eg: **49,999** = 4 by just ignoring all the 9s.

When a large number has several digits that add up to 9, they can also be cast out:
eg: **786,432** has 7+2=9 and 6+3=9 leaving 8+4=12=3

nb: When electricity is applied to Quartz Crystal, it gives off a specific frequency of 786,432 cycles per second or Hertz. This quartz electro-stimulation frequency has been selected from Jain's immense dictionary of numbers called **"Harmonic Stairway"** which is currently being made available as an illustrated online dictionary for mathematical and scientific researchers. Thus this number 786,432 is a special number that we call Anointed as it is not an invented or arbitrary number, it has special qualities. The aim of this encyclopedic compendium is that a student one day may access some of these qualities and take the knowledge further.

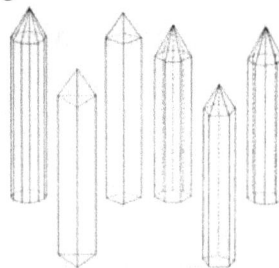

(image from book: "The Crystal Connection" 1987 courtesy of Randall and Vicki Baer)

Some modern day mathematicians think that this concept of Digital Compression of Big Numbers Being Reduced to Single Digits is a form of Numerology that has no value; it is sad that they can't get beyond this limitation, and realize that it truly opens doors or windows to Higher Number Awareness and Atomic Art.

As a brief exercise, to show that you understand how to convert Base 10 Numbers into Base 2, (meaning the student needs to know all their powers of 2), complete the following box or chart that was explained 4 pages earlier on how to convert the number 108 in Base 10 into its Base 2 expression:

In the 1st row below, are the Powers of 2.

In the 2nd row, write out their values eg: $2^2=4$.

In the 3rd row, mark the symbol "✓" where a Power of Two has been used.

In the 4th row, where a "✓" has been used, replace it with the numeral "1" to indicate that it is a "Yes", and where there is no tick, or no Power of 2 used, fill it that cell with a "0" or Zero to indicate that it is a "No".

108							
2^7	2^6	2^5	2^4	2^3	2^2	2^1	2^0

Write out the Base 2 Expression: ....

ANSWER: Mathematically, it is written as:

108 $_{10}$ = 1101100 $_2$ where the subscripts indicate the bases used.

THE ART OF NUMBER

THE TRANSLATION OF NUMBER INTO
ART

THE JAIN 108 COSMOMETRIES
OR 9 CELESTIAL TRANSCIPTS

COMPRESSION

OF THE
MULTIPLICATION TABLE

+

THE 3 DISTINCT PHI CODES
OR THE 3 DIALS

The **9**

CELESTIAL TRANSCRIPT CODINGS

(CTC)

by JAIN 108

Here is a listing of the 9 most essential Visual/Artistic Codes derived purely from Mathematics. Most of these listings will be covered in this book.

• CTC 1 - The Tessellatted Magic Square of 3x3 Diamond.
— The **Magic Square Of Six by Six**, that adds up to 666, whose pattern forms the **Swaztika**, the original solar symbol. To go into the next galaxy, you had to go through the Eye Of the Sun, (This solar swaztika symbol has been demonized to create fear which denies access to inter-dimensional travel). We are Fearless.

• CTC 2 - The Compressed Multiplication Table
— The **4 Times Multiplication Table Compressed** forming Octagram
— "**All The Ones**" forming the **atomic structure of Rutile.**
— The superimposition of "**All The Ones**" + "**All The Eights**" forming the **Atomic Structure of Platinum Crystal.**

• CTC 3 – The True Value of Pi (3.144... known as Jain Pi) **based on the Square Root of Phi (1.272...). Traditional Pi is a Lie, and is Disharmonic.**

• CTC 4 – The **Compression of the Binary Sequence** forming the VW **Symbol** and shadow of the CubOctahedron (aka CubeOcta).
— The Geometric Combination of The Binary Code with the Phi Code, based on Concentric Circles, to show that there is no separation, and the Phi Ratio is in all the Geometries of Creation and Light Harmonics.

• CTC 5 – The **Prime Number Cross** formed from the **Wheel of 24,** forming the **4th-Dimensional Templar Cross.**
— **Ulam's Rose** plotted from the spiralling natural progression of counting numbers   .

• CTC 6 – You are now a candidate of Rapid Mental Calculation which increases your Memory Power and Mathematical Confidence.

• CTC 7 –The infinite dance between the Icosa-Doceca, how they stellate and travel from the atom to the universe, principle of infinite expansion and contraction.
— The **Cuboctahedron** or Jitterbug, the pure principle of **Shape-Shifting,** how this Archimedean Solid creates the 5 Platonic Solids.

• CTC 8 – The **Conversion of the 1-Dim Fibonacci Sequence** to form the 2-Dimensional Phi Spiral, 3-Dim Ram's Horn Phi Spiral and 4-Dim self-organizing Torus.
— The **Language of Light** creation of the various Earthly and Celestial **Alphabets** based on the 3-Dimensional Phi Spiral indexed by intelligent tilt angles of the tetrahedron, dodecahedron and the torus.

- CTC 9 –The 3 Phi Codes or The 3 Dials:

The Mystery of the 108 Code in the Divine Proportion is really based on a hidden Trinity. There are actually 3 Phi Codes, each of which sum to 108. All 3 Codes are based on 24 infinitely repeating digitally compressed numbers. When the string of 24 digitally compressed digits are rewritten as a 3x8 rectangular array, they form the ancient Solfeggio Scales triplets of the phone pad like 1-4-7 and 2-5-8 and 3-6-9 and anagrams thereof. Have a look at the table below.

— Phi Code 1: The **Compression of the Fibonacci Sequence** to form the infinitely repeating 24 Pattern whose code is known as **Shri 108** (12 Pairs of 9 = 108). The Linear Sequence expressed as: **PC1-(1,1,2)** short for:
1, 1, 2, 3, 5, 8, 4, 3, 7, 1, 8, 9, 8, 8, 7, 6, 4, 1, 5, 6, 2, 8, 1, 9

— Phi Code 2: — **The Powers Of Phi 108** forming another unique 108 Pattern: the multi-dimensional Jacob's Ladder known as The Caduceus Symbol (and applied by Big Pharmaceutical Companies). It is approximated by the Lucas Sequence:
1 - 3 - 4 - 7 - 11 - 18 - 29 - 47 - 76 -123
It is expressed as: **PC2-(1,3,4)** short for:
1, 3, 4, 7, 2, 9, 2, 2, 4, 6, 1, 7, 8, 6, 5, 2, 7, 9, 7, 7, 5, 3, 8, 2

— Phi Code 3:
It is expressed as: **PC3-(1,4,5)** short for:
1, 4, 5, 9, 5, 5, 1, 6, 7, 4, 2, 6, 8, 5, 4, 9, 4, 4, 8, 3, 2, 5, 7, 3

All 3 Phi Code Dials surprisingly form the ancient musical Solfeggio Scale triplets:

1	1	2	3	5	8	4	3
7	1	8	9	8	8	7	6
4	1	5	6	2	8	1	9

(nb: this rare and never before published information will be released in my next book: THE BOOK OF PHI, volume 5).

~ CHAPTER 1 ~

THE BINARY - DOUBLING SEQUENCE

The MAGIC Of NINE

HIDDEN WITHIN The DOUBLING SEQUENCE

The MATHEMATICAL And GEOMETRICAL ORIGIN Of The VW SYMBOL

CHAPTER CONTENTS

PART 1

Here is a typical **JAIN MATHEMAGICS WORKSHEET** designed for Teenagers. I find that it is more fruitful to encourage a student to discover things for themselves. By just giving a student all the final data is not as powerful as realizing Laws of Numbers for yourself.

You can complete this table below. Ultimately you will want to know as many Powers of 2 as is possible for you, as this stimulates your Memory Power and therefore your mathematical confidence.

Exercise 1

On this page, in the space below, write down The Doubling Sequence, from memory, that is, as far as you can go. Such exercises help you develop Memory Power.

It starts with:

1 - 2 - 4 - 8 - 16 -

On the following page, complete the table up to "2 to the Power of 17"

aka 2^17

aka 2^{17}.

Then observe for any repeatability of data.

JAIN MATHEMAGICS FOR TEENS: WORKSHEET
(The Doubling Sequence)

2^n Powers of 2	DOUBLING SEQUENCE	ADD DIGITS	REDUCED SUM to a single digit
2^0	1	1	1
2^1	2	2	2
2^2	4	4	4
2^3	8	8	8
2^4	16	1+6 =	7
2^5			
2^6			
2^7			
2^8			
2^9			
2^{10}			
2^{11}			
2^{12}			
2^{13}			
2^{14}			
2^{15}			
2^{16}			
2^{17}			

a) - Write down the numbers that display Recursion or Repetition:

....

b) - What is the Periodicity of this Doubling Sequence? ....

c) - What happens when you cut this Sequence of 6 infinitely repeating digits is cut in half, and viewed as 2 rows of 3 digits?

....

d) - Can you remember all these numbers of this Doubling Sequence? Improve your memory power, and therefore your confidence, by learning this sequence as far as it is comfortable to do so.

The **"Powers of 2"** is the **Doubling Sequence:**

a) 1, 2, 4, 8, 16, 32, 64, 128, 256, 512, 1024, 2,048, 4,096, 8,192, 16,384, 32,768, 65,536, 131,072

Here is a list or table for all the **Powers of 2** up to "2 to the Power of 17"

aka 2^17   aka 2^{17}.

We then observe for any **repeatability** or **recursion** of data. That which "repeats infinitely" gives us a handle on understanding "the Mathematics of Infinity".

POWERS OF 2^N	DOUBLING SEQUENCE	ADDING ALL THE DIGITS	REDUCED SUM TO A SINGLE DIGIT
2^0	1		= 1
2^1	2		= 2
2^2	4		= 4
2^3	8		= 8
2^4	16	= 1+6	= 7
2^5	32	= 3+2	= 5
2^6	64	= 6+4	= 1
2^7	128	= 1+2+8	= 2
2^8	256	= 2+5+6	= 4
2^9	512	= 5+1+2	= 8
2^{10}	1,024	= 1+0+2+4	= 7
2^{11}	2,048	= 2+0+4+8	= 5
2^{12}	4,096	= 4+0+9+6	= 1
2^{13}	8,192	= 8+1+9+2	= 2
2^{14}	16,384	= 1+6+3+8+4	= 4
2^{15}	32,768	= 3+2+7+6+8	= 8
2^{16}	65,536	= 6+5+5+3+6	= 7
2^{17}	131,072	= 1+3+1+0+7+2	= 5

Fig 1

Table or Chart of the Doubling Sequence up to 2^17

and listing the Digital Compressions or Sums thereof.

Answers 3:

a) – By studying the data of Digital Compression on the furthermost column on the right hand side, there appears an obvious repetition or recursion of 6 reduced single digits:

1   2   4   8   7   5

b) - This is technically called a **Periodicity of 6**:

c) – When this Sequence is cut in half, two horizontal rows of 3 digits appear in this form.

1   2   4

8   7   5

Keep looking at these 6 digits until the instantaneous magic of Pattern Recognition comes like a flash of lightning into your Consciousness.

Can you see that the **3 vertical pairings have sums of 9**:

$1 + 8 = 9$ $2 + 7 = 9$ $4 + 5 = 9$

1   2   4

8   7   5

_____

9   9   9

To improve your **Memory Power** and **Confidence** it is a good practice to learn the Doubling Sequence as far as you can manage, at least up to 2^{14}.

Did you notice that the numbers missing in this sequence are **3-6-9** which form an equilateral triangle in the 9-Point Circle!

PLOTTING The INFINITELY RECURRING PATTERN
Of The DOUBLING SEQUENCE
UPON The 9–POINT CIRCLE

Exercise 4

Plot the Doubling Sequence: **1 - 2 - 4 - 8 - 7 - 5** on the 9 Point Circle, by joining a long, unbroken, continuous line that connects 1 to 2, then 2 to 4, then 4 to 8, then 8 to 7 then 7 to 5. To keep the electrons flowing (as if this is a copper circuit) it is a good practice to join the last number "5" back to the first number "1", a reminder of the biblical "Alpha and the Omega" the First and the Last.

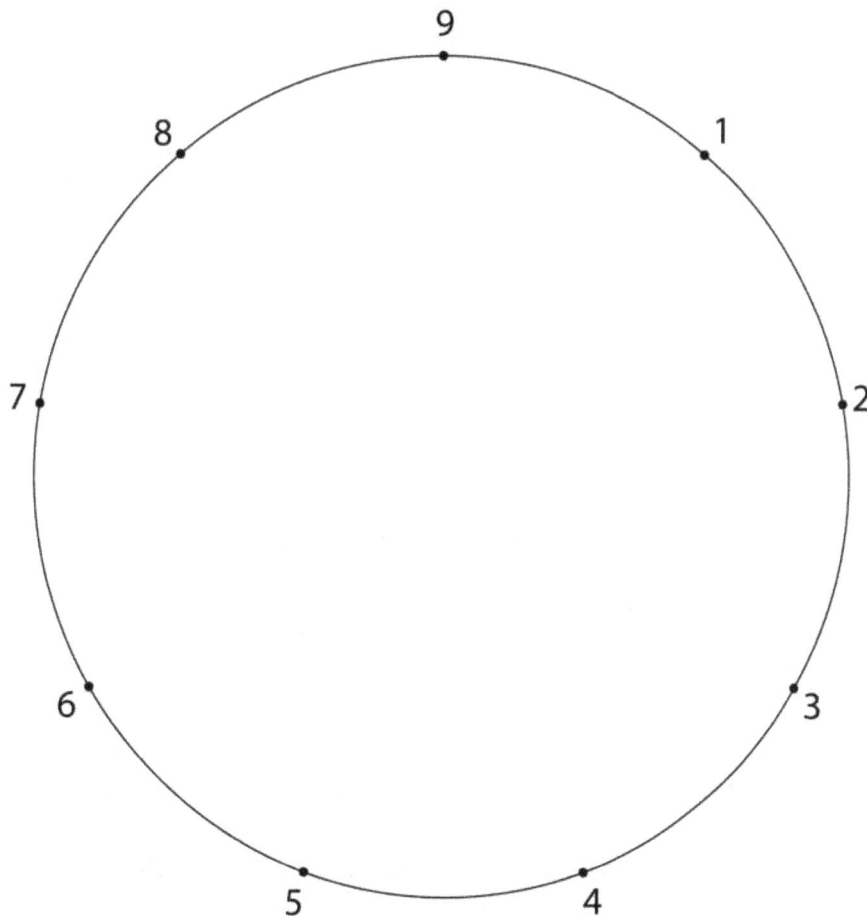

Fig 2

The 9-Point Circle

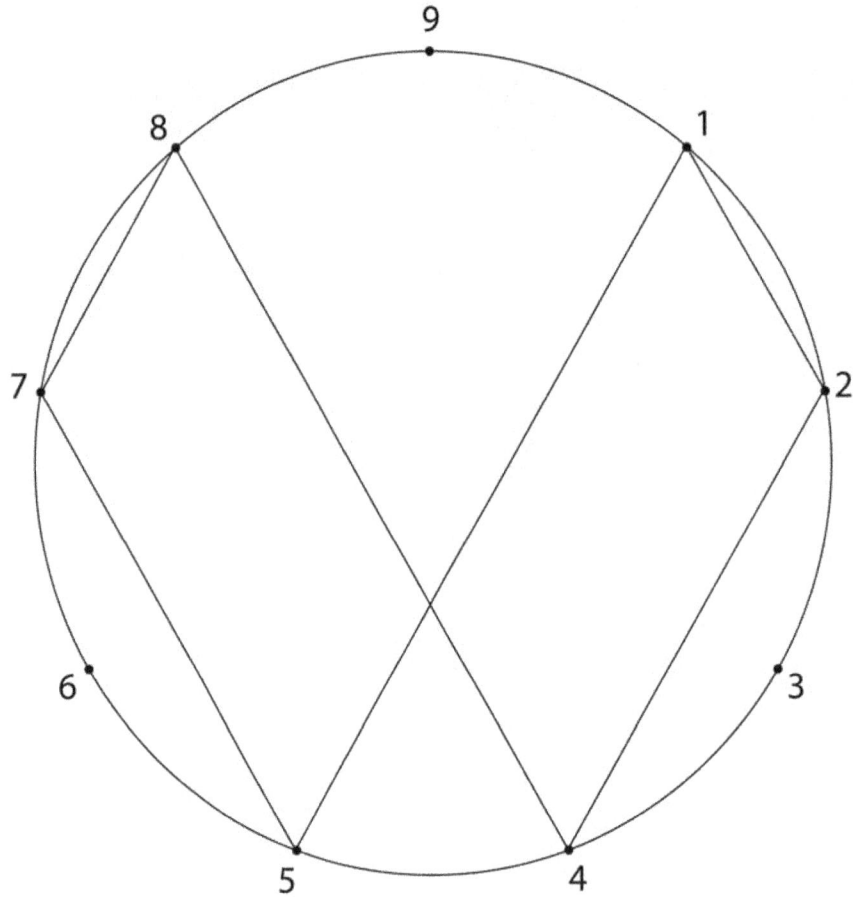

Fig 3

The Digitally Compressed Binary Pattern

upon the 9 Point Circle connecting the infinitely

repeating pattern of 1-2-4-8-7-5.

Notice that there exists an obvious

Mirror Axis or Symmetry of Reflection

of the Doubling Sequence:

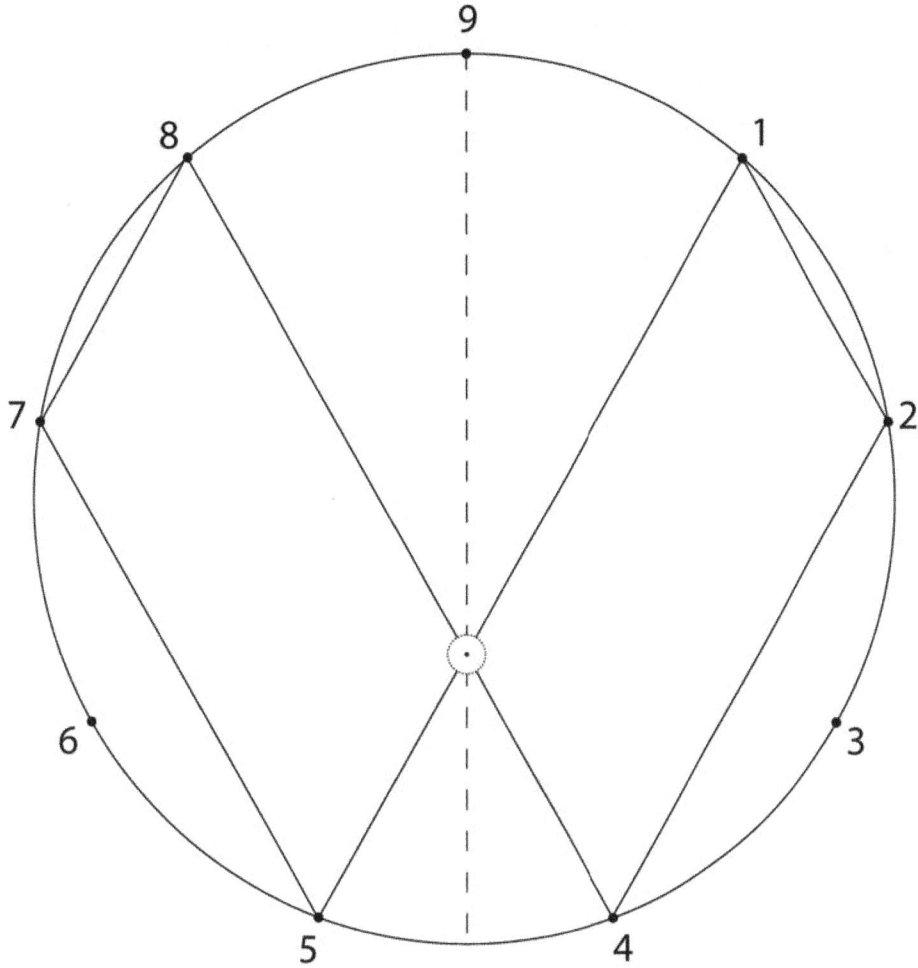

Fig 4

The Digitally Compressed Binary Code

exhibiting a mirror-axis (in dotted lines) through

the Number 9 and the midpoint of 4 & 5.

Notice also that the centre of gravity

is not the actual centre of the circle, but is "off-Centre",

a distinct nexus point where the 2 major lines intersect

and is known as The Emanation Point.

When a coloured pencil or high-light marker is traced over some of the lines just constructed, a well-known commercial logo is generated from this design? This subject leads to The **MATHEMATICAL ORIGIN Of SYMBOLS And COMMERCIAL LOGOS.**

Let us highlight some of these lines:

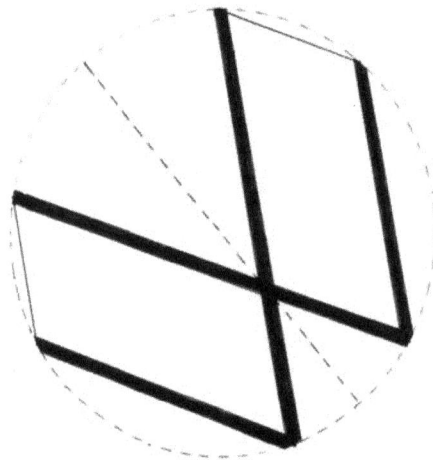

Fig 5

Highlighting some lines of the Digitally Compressed Binary Code to suggest the formation of a famous symbol.

Notice where the Line of Symmetry is, running along from the 9 Point to the midpoint of 4 and 5. If you to view this diagram as an origami exercise, you will notice that this dotted line is an axis of symmetry, found by folding the pattern over itself and checking to see that all lines overlap exactly or superimpose upon the other half. **Symmetry** is precisely what the Pattern Hunter is looking for.

Fig 6

The VW Symbol, imbued in the mass consciousness

The Logo of Mathematical Origin that was consciously derived from this Doubling Sequence In The 9 Point Circle is the **VW** symbol. ("The Volks Wagon" or People's Car, designed by Hitler. Most world leaders use sacred symbols of the highest mathematical order as if to claim this secretly potent Universal Force).

What would this pattern look like, when it is drawn upon itself 6 times, at the same 40° angle each time? (since the 9 Point Circle has its 9 vertices at every 40° angle, 9x40=360°).

When I am using the word "**Code**" I am distinguishing it from the word "**Sequence**".
The "Binary Sequence" has multiple digits, like:
1, 2, 4, 8, 16, 32, 64, 128, 256, 512, 1024,
whereas the Code refers to the Digital Compression of such a sequence, and can only contain recursive single digits from 1 to 9.
(The same applies to the **Fibonacci Sequence** of 1-1-2-3-5-8-13-21-34 etc but the **Phi Code** has 24 recursive **single** digits).

Fig 6a

The VW Symbol, imbued in the mass consciousness

ROTATING THE BINARY CODE PATTERN (VW) SIX TIMES TO GENERATE
THE ENNEAGRAM OR 9 POINTED STAR SYMBOL

The Binary Sequence Digitally Compressed [1-2-4-8-7-5]
plotted upon the 9 Point Circle 6 times @ increments
of 40°. Its the same pattern for 9 x as the lines repeat.

[DECAL DESIGN] for Amplification of
Wealth, Health. Its Energy is to Double continuosly,
so be careful you do not magnify any negative traits!

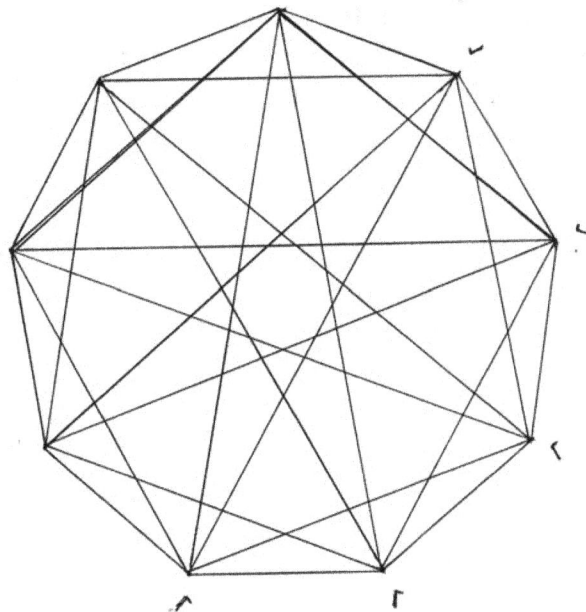

This Yantram (Power Art) is a partial (semi-complete)
Mystic Rose of 9 diagram / Enneagram. It has a signature
of 9-9-9-9 ie 4 Enneagrams bounded by the inner and outer Enneagons.
[Doubling Sequence : 1 - 2 - 4 - 8 - 16 - 32 - 64 - 128 - 256 - 512 → ∞]

1	2	4
8	7	5

3 Pairs of 9.

by JAIN 108
15-8-2010 mullumbimby.

Fig 7

**A typical sketch by Jain translating Number Into Art,
experimenting with the idea of multiple rotations.
The element of surprise comes from not knowing
the final outcome of the design.
Creation of the Enneagram.**

Here is the same pattern again for purposes of reviewing and colouring in:

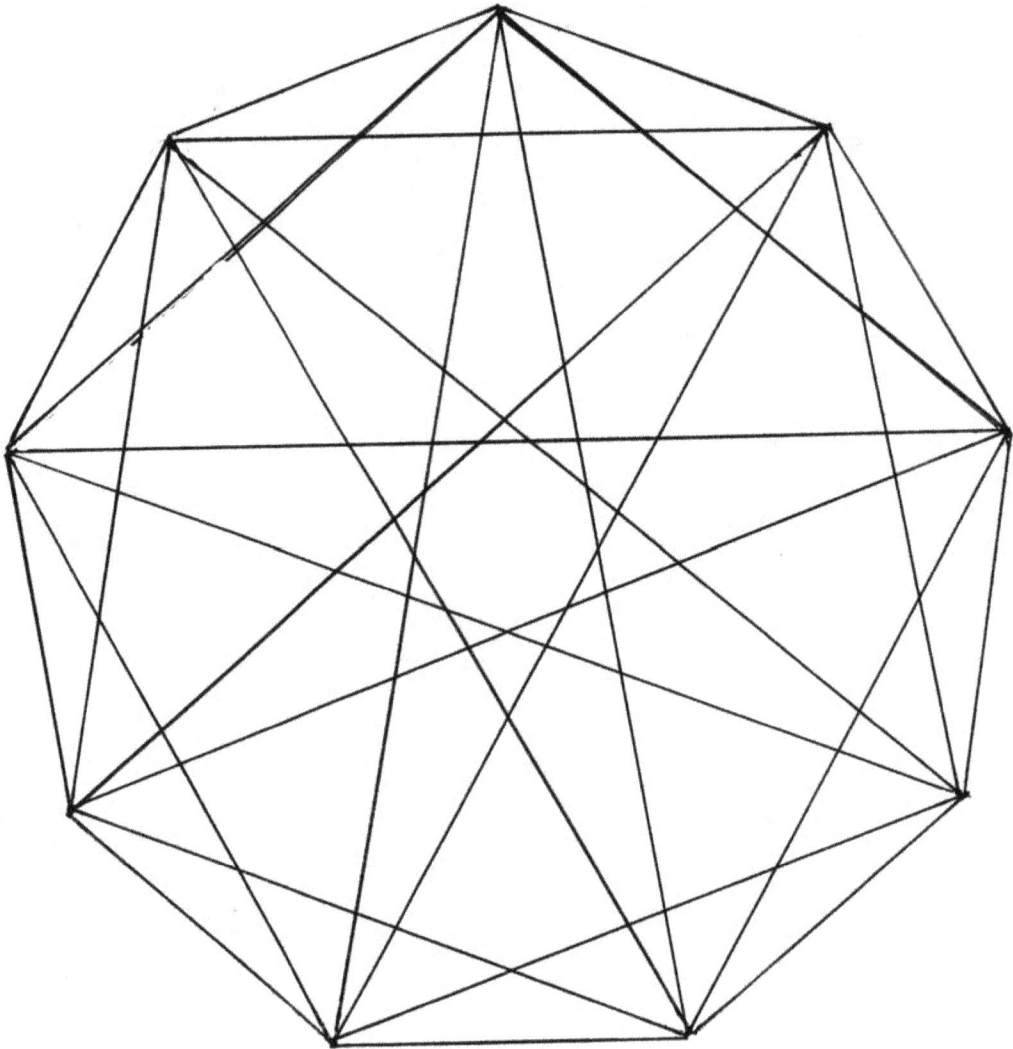

Fig 7a

The Digitally Compressed Binary Code Yantra 1-2-4-8-7-5

plotted upon the 9-Point Circle at 6x40° forming the Enneagram.

DID YOU KNOW that this 2-Dimensional outline of **VW** is the **shadow** of a hyper-dimensional Archimedean Solid called the **Cuboctahedron**!

Have a look at this 3-Dimensional form of the Cuboctahedron with squinted eyes, as if to look through the form, and check that you can see the VW outline in its 2-dimensional form:

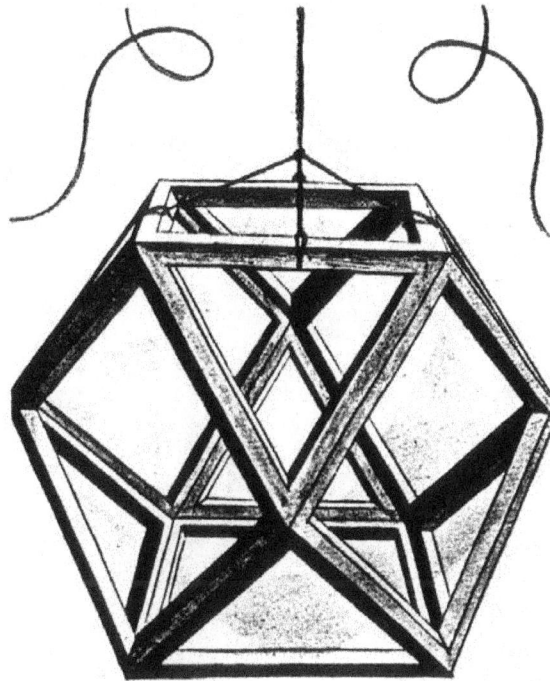

Fig 8
The Cuboctahedron as drawn by Leonardo da Vinci is the multi-dimensional form of the flatland VW symbol.

Leonardo da Vinci's drawing of a Cuboctahedron (Fig 8 above) is formed from the **12 centres of 12 kissing spheres around a central sphere** producing 6 squares and 8 triangles when the vertices or points are joined.

It is the most fractal shape in the universe, as its internal vectors or radii lengths emanating from the centre as the same length as are its outer edge lengths, as shown in Fig 9.

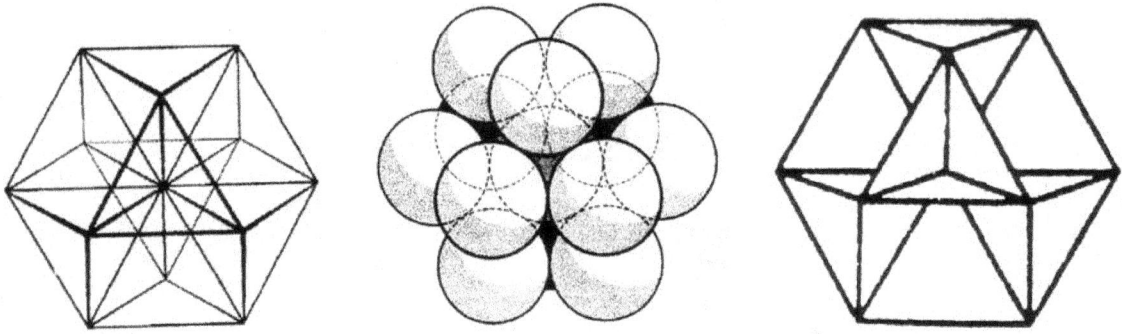

Fig 9

The inside of the Cuboctahedron has 24 triangular vectors, thus linking it to Nature's Phi Code of 24 Repeating Numbers.

This is the ultimate definition of Fractality, the opposite of Fractionating, because the Inside is the same as the Outside!

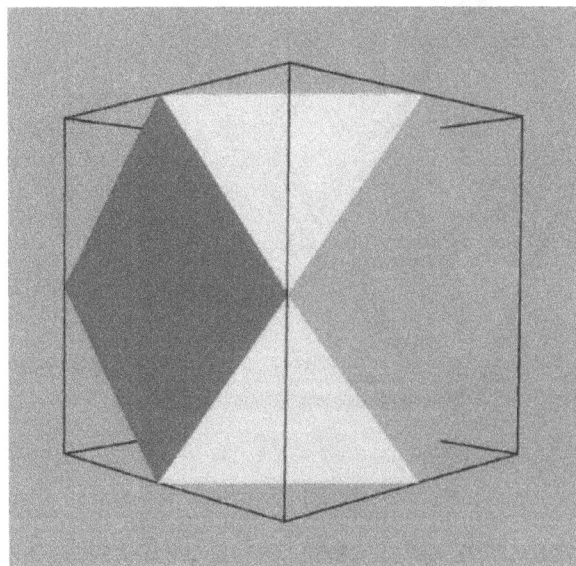

Fig 10

Can you see the VW symbol in the Hexagonal view or shadow of the 3-Dim Cuboctahedron?

PART 2
FUSION or MARRIAGE of the CIRCULARIZED PHI SPIRAL & BINARY CODE
"Validating the Art of Digital Compression"

THIS IS THE MEETING OF TWO WORLDS:

The Biological series of Nature and the Hi-Tech Borg Binary Codes, yet they are One Inseparable, their Marriage dispels the trap or illusion of Duality or Separation from God. Their Union or combination mysteriously aligns their two individual Centres or Eyes almost exactly upon one another, to form One common centre, One Eye!

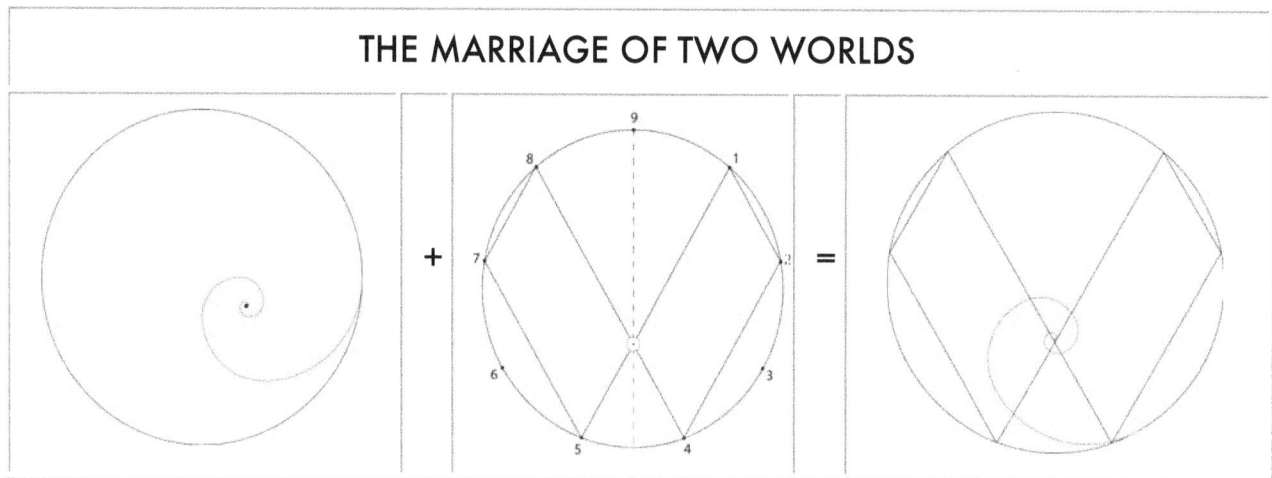

THE MARRIAGE OF TWO WORLDS

HOW TO CREATE THIS FUSED SYMBOL:

First, turn the 9 Point Circle 40° to the right, so that the Number 9 is at the zenith or top most northern point, as shown below in Fig 11 below.

Observe this Nodal Point, Centre Point, that "Sweet Spot" where the lines cross upon the axis of folding.

Another Pattern Hunter Marko Rodin (www.markorodin.com of Vortex Based Mathematics) focusses deeply upon this **"Emanation Point"** as he terms it, and creates a whole cosmology upon this.

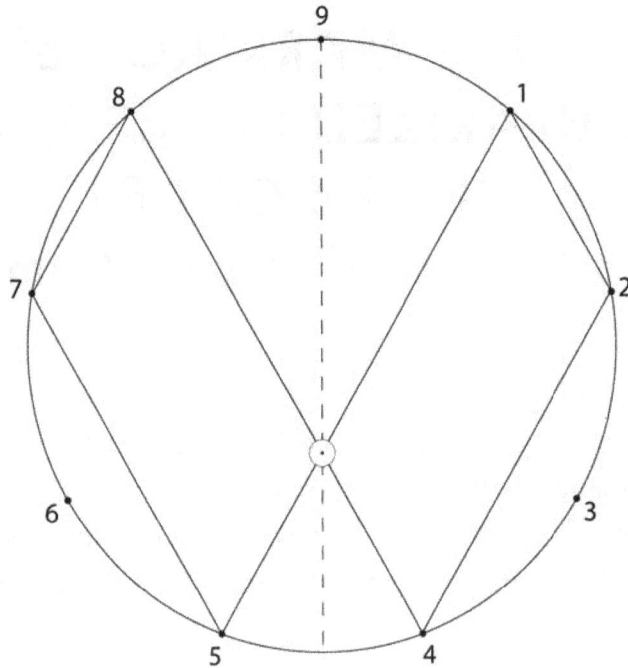

Fig 11

The Emanation Point of the Binary Code in the 9-Point Circle, as cosmologized by Marko Rodin

This is to show you that many other numerical nomads or mathematical monks like myself, have spent their lifetime investigating such Symmetry.

Rodin's colleague Nassim Haramein of The Resonance Project (http://theresonanceproject.org/ a distinguished physicist rewriting all of Einstein's equations in terms of the Golden Mean 1:1.618033... and sees everything from atoms to humans to galaxies as tori) decided to overlay or superimpose upon this pattern a circularized form of the Golden mean Phi Spiral, (already drawn up by Nabokov, hence known as the Nabokov Spiral) and noticed that the 2 centres of both diagrams line up or overlap almost precisely.

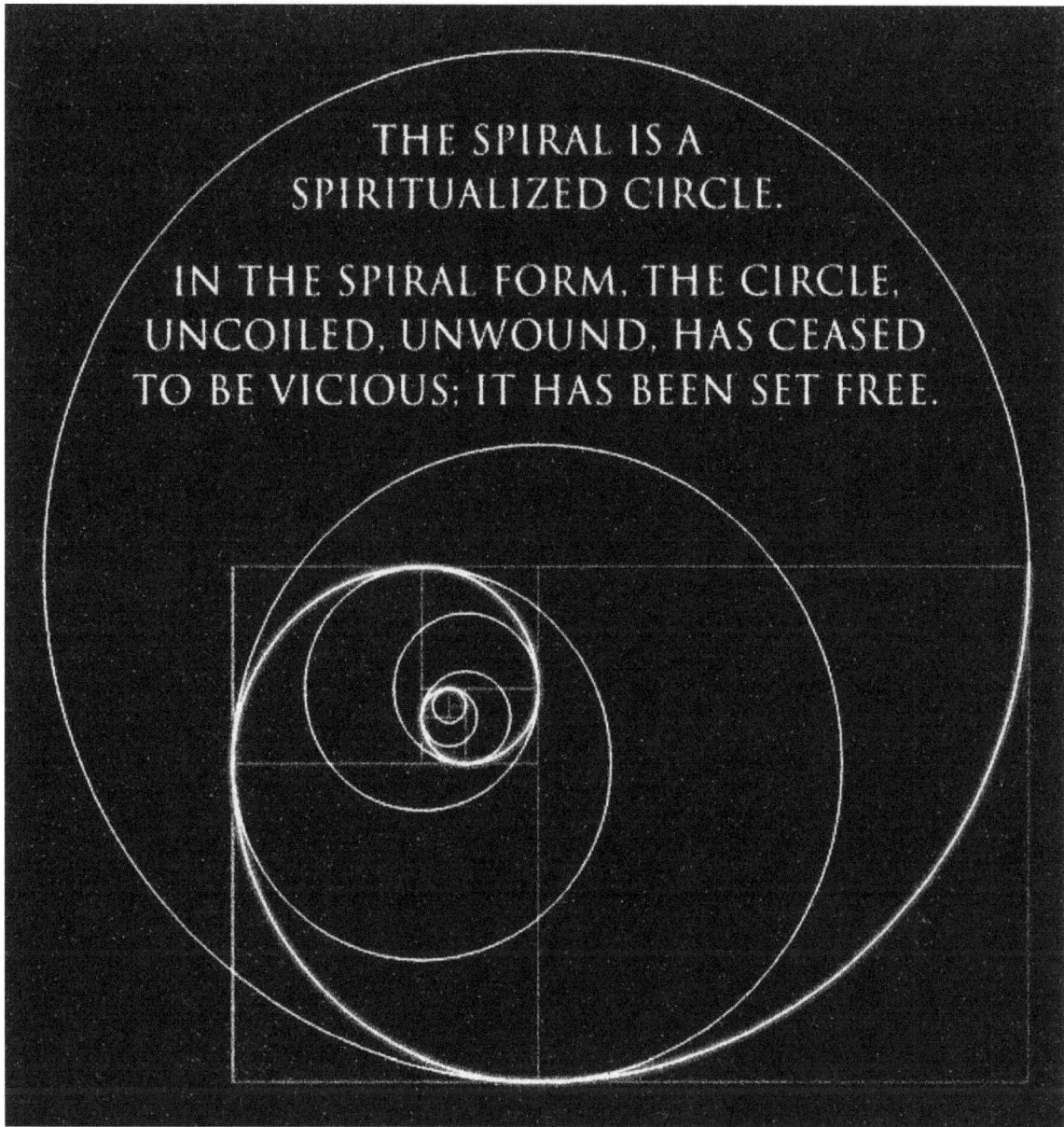

THE SPIRAL IS A
SPIRITUALIZED CIRCLE.

IN THE SPIRAL FORM, THE CIRCLE,
UNCOILED, UNWOUND, HAS CEASED
TO BE VICIOUS: IT HAS BEEN SET FREE.

Fig 12

Nabokov Spiral is the Phi Spiral Circularized.

I have redrawn this:

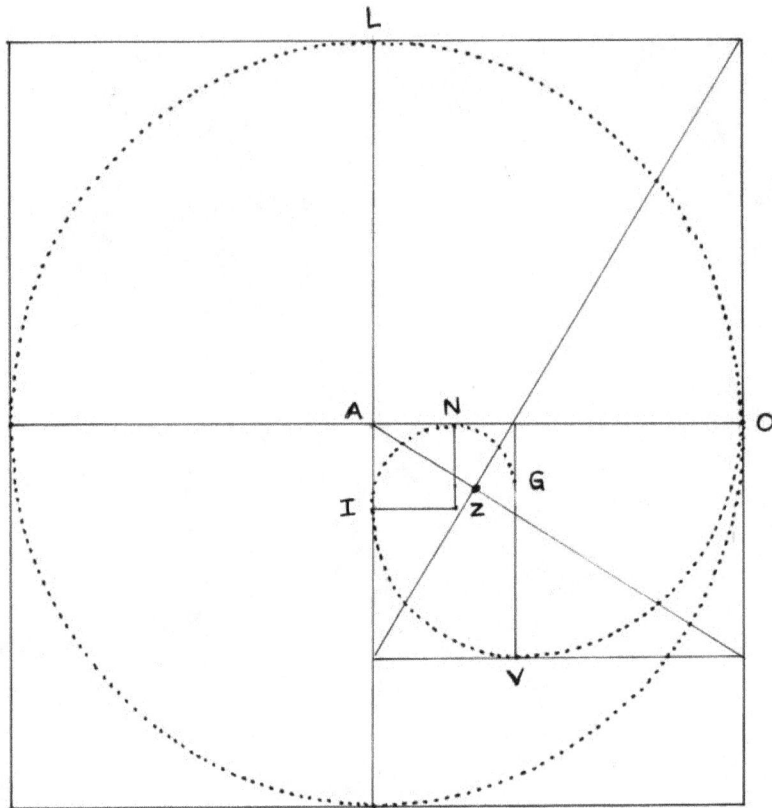

(The Centre of the Square-Circle is point "A"
(The Centre or Eye of the Phi Spiral is "Z"
nb: This Golden Mean Centre "Z" by definition is "off-centre". This composition of
Phi Spiral, when circularized or framed in
a Square, is called the Nabokov Spiral.
(The Path of the Phi Spiral can be seen
as the 5 $\frac{1}{4}$ circle arcs, expressed above as
the Path of "L-O-V-I-N-G" JAIN.

Fig 12a

**Jain's hand-drawn rendition of the Nabokov Spiral,
showing the quadrated circle that forms
the circularized phi spiral.**

Here is The Circularized Phi Spiral:

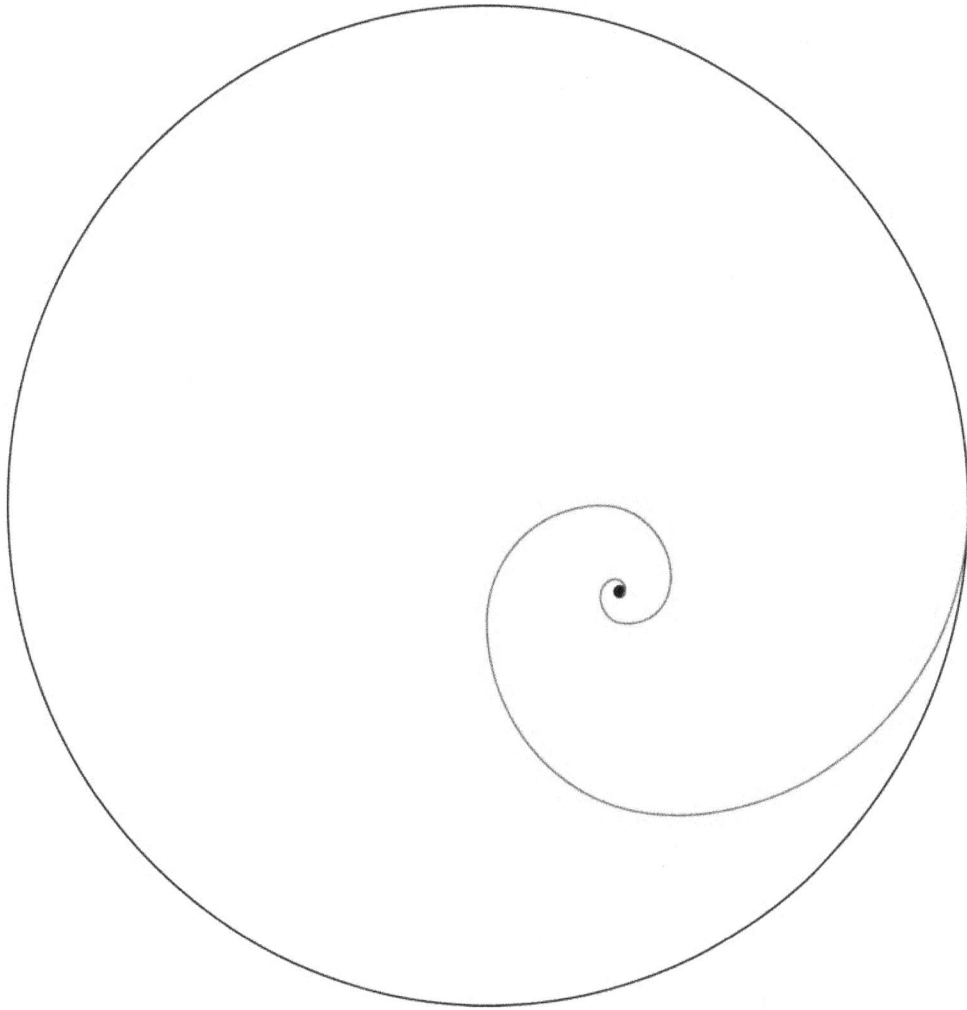

Fig 12b

**Jain's computerized rendition of the Nabokov Spiral,
showing only the circularized phi spiral.
The dot shown, as expected, is off-centred,
it is the actual eye of the Spiral, the Tornado,
the Sweet Spot.**

If you look at lily's in a pond, where the stalk meets the flat floating leaf, the stalk never meets the centre of the circle, it meets this Sweet Spot, the sinkhole, the wormhole.

Here is a rather complex sketch, in Fig 13 below, of the two entities that concludes with the fact that the two centres are almost perfectly in synchronization.

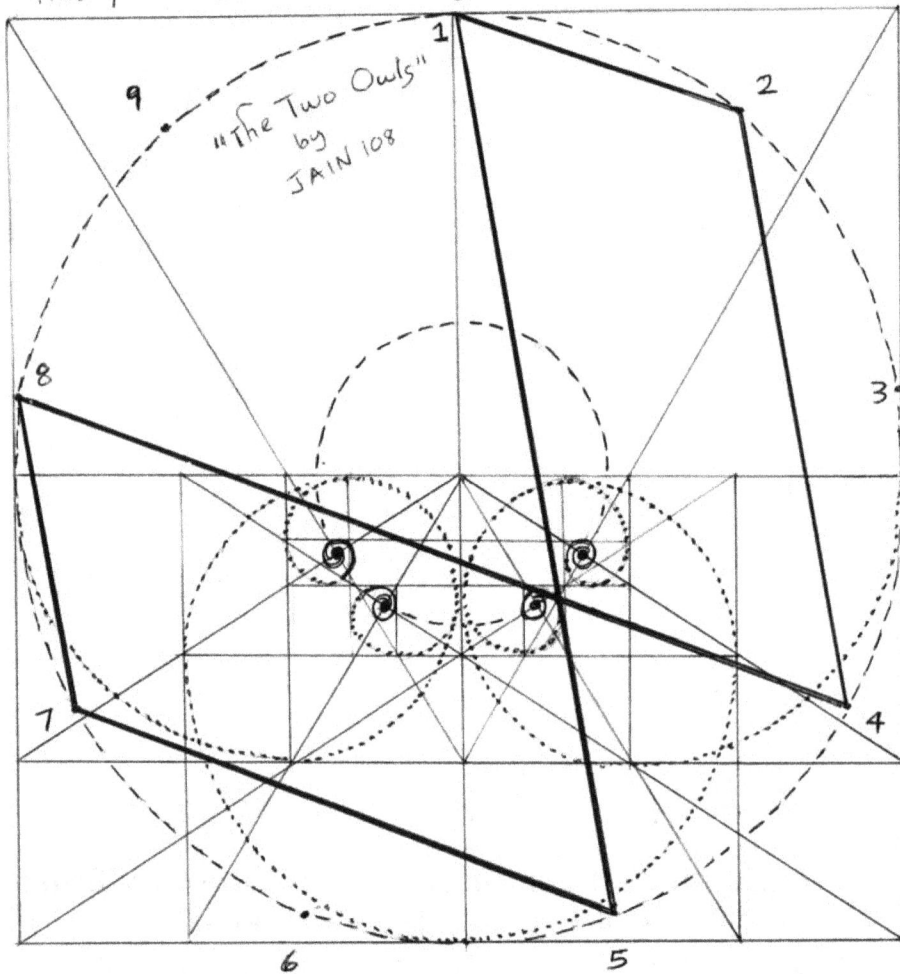

SuperImposing 4 Phi Spirals and Binary VW symbol to see if their Centres SuperImpose. Answer = YES. The 4 EYES of the φ Spiral Cent͟ s lie on the Inner Cir that passes thru Nexus or Eye of Binary VW Symbol.

"The Two Owls"
by
JAIN 108

What is radius of Inner Circle? if larger Circle has Radius 100mm
≐ 1/3 ≐ 33.3 mm, This is the "black hole" Geometry of the Void. This Diagram validates Digital Compression as a true Science : It expresses the Binary Code of 1-2-4-8-7-5 as a Uni-Phi Field.   JAIN   6-3-2010

Fig 13

4 Phi Spirals, 2 owls, 1 Binary Code, all lining up.

Two distinct worlds meeting in their Centres.

Since the Phi Spiral Vortex is a wormhole, it defines a broadband, so no exact number can be allocated to its centre, since the phi ratio 1:1.618033... has its decimal form travelling forever. This means the 2 centres can never touch, but approximate.

This is an important discovery, realized by merely Joining The Dots. This is all that I have done in the last 30 years, just joining the dots.
I believe that these two geometries are like two worlds that bridge or come together, the world of Phi 1-1-2-3-5-8-13, and the world of Binary 1-2-4-8-16-32-64. They actually meet in their centres...
Phi is based on 1 becomes 1 becomes 2 becomes 3 becomes 5 etc and binary is 1 becomes 2 becomes 4 becomes 8 becomes 16 etc...
One is the world of Nature and Biology, and the Other is the world of Computers and Digital Creations... thus, we are both codes!
This also **validates the supreme Art of Digital Compression**, that is the thesis of my life's work called The Translation of Number Into Art or simply **THE ART OF NUMBER**, that it is inextricably related to Atomic Art, Nuclear Geometries and the Fabric of Creation.

Fig 13a below shows the **Union of the Circularized Phi spiral and the VW Binary Code** in a simpler and clearer format.
It is the coming together of two worlds: the Natural and the Alien.
1 - the Circularized Phi Code is the world of Biology and Living Mathematics of Nature, whereas the

2 - Binary Code / Doubling Sequence is the world of crystals / silicon chip molecules / high tech and computers and machines, the space age and Lords of the electrons...

Mathematics teaches us that all worlds are connected. Our world of Nature and Biology is integrating its understanding of the rapidly advancing world of technology and the coming age of intelligent robots. Consciousness is also in metals and crystals, this mathematics is teaching us that all life forms are possible in this toroidal universe.

Below, I have combined the two patterns: The Circularized Phi Spiral and the Digitally Compressed Binary Code In The 9-Point Circle.

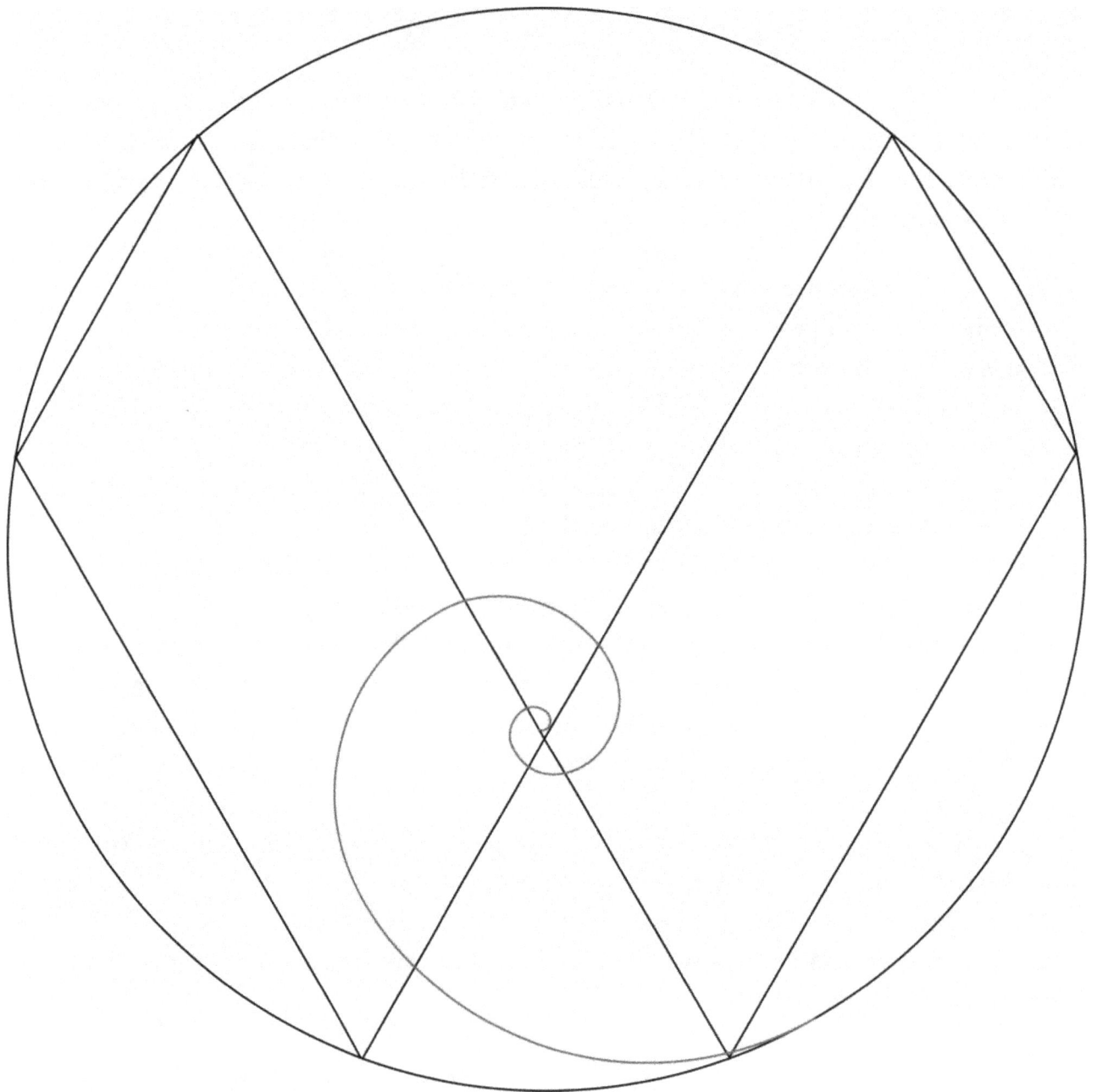

Fig 13a

The Fusion or Union or Marriage of
the Circularized Phi Spiral and the Binary Code VW,
sharing a common central Eye or Wormhole,
the key perhaps to the Nuclear Physics of Black Holes
and the ancient Art of Time Bending.

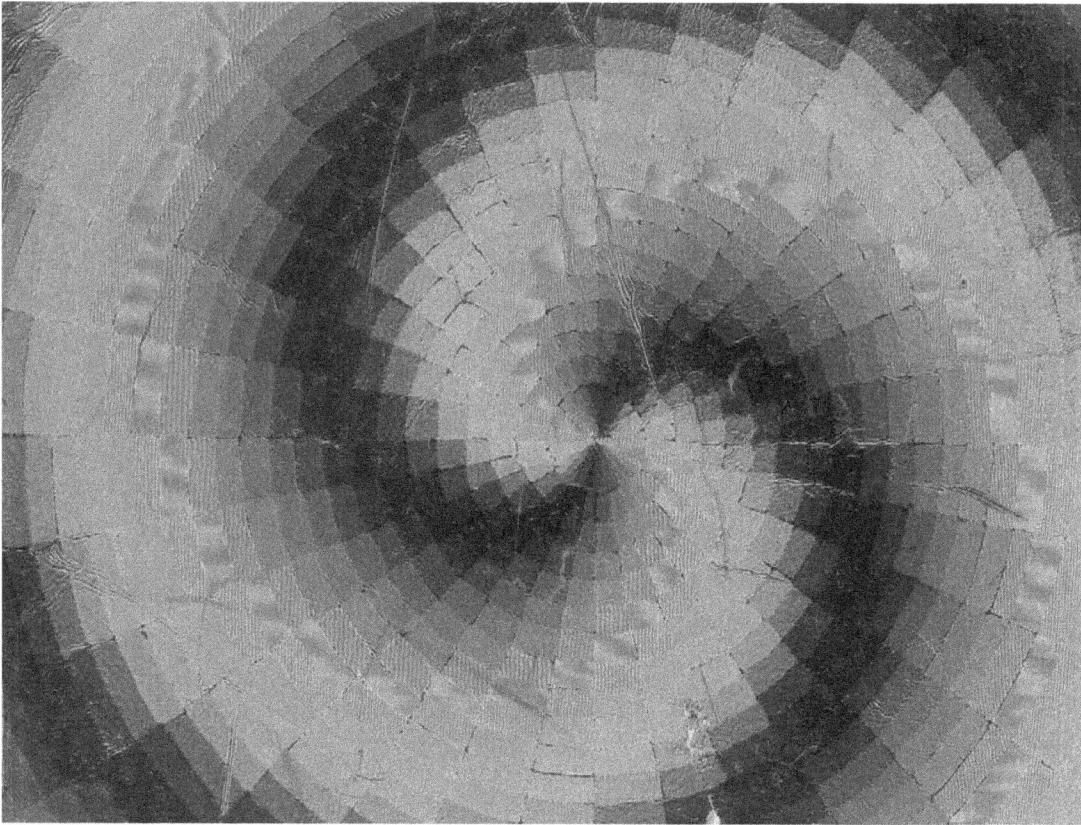

[footnote:
(If you google "Phi Spiral Circle" under Images, you will come up with this
Nabokov Spiral:
http://spiralzoom.com/Culture/ContemporaryArt/spiralnabokov.jpg
The writing says: "The Spiral is a spiritualized Circle. In the spiral form, the circle
uncoiled, unwound, has ceased to be vicious: it has been set free".
image taken from: – Vladimir Nabokov, Speak, Memory (1966), author of Lolita).]

~ CHAPTER 2 ~

THE ODD NUMBERS GENERATING THE SQUARED & CUBIC NUMBERS

PART 1
UNIVERSAL LANGUAGE of PATTERN RECOGNITION

Examine the common SHAPE of the Answers to the following Sums of the Odd

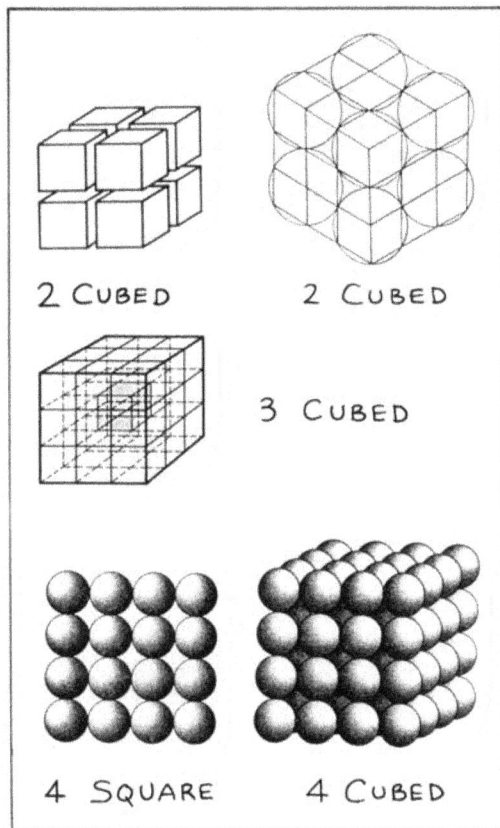

2 CUBED   2 CUBED

3 CUBED

4 SQUARE   4 CUBED

Number Sequence 1-3-5-7-9-etc as shown in Fig 1.

By knowing this shape, a student can mentally predict the answers, by filling in the box below to the questions in rows c, d, e, f, g, h, i, j, k, l and m.

What is your Conclusion?

Fig 1, without the answers in the right hand column, is a typical "Jain Mathemagics Worksheet" which allows the teacher or parent to photocopy this and challenge the student or child to discover inherent or underlying symmetry and patterning.

a	1	= 1	= 1x1	= 1^2
b	1 + 3	= 4	= 2x2	= 2^2
c	1 + 3 + 5	=	=	=
d	1 + 3 + 5 + 7	=	=	=
e	1 + 3 + 5 + 7 + 9	=	=	=
f	1 + 3+ 5 + 7 + 9 + 11	=	=	=
g	1 + 3 + 5 + 7 + 9 + 11 + 13	=	=	=
h	1 + 3 + 5 + 7 + 9 + 11 + 13 + 15	=	=	=
i	1 + ... + 17	=	=	=
j	1 +. . . . . + 19	=	=	=
k	1 +. . . . . + 21	=	=	=
l	1 +. . . . . + 23	=	=	=

Fig 1

Typical Jain Mathemagics Worksheet to discover the pattern formed
when consecutive odd numbers are progressively summed.

a	1	= 1	= 1x1	= 1^2
b	1 + 3	= 4	= 2x2	= 2^2
c	1 + 3 + 5	= 9	= 3x3	= 3^2
d	1 + 3 + 5 + 7	= 16	= 4x4	= 4^2
e	1 + 3 + 5 + 7 + 9	= 25	= 5x5	= 5^2
f	1 + 3+ 5 + 7 + 9 + 11	= 36	= 6x6	= 6^2
g	1 + 3 + 5 + 7 + 9 + 11 + 13	= 49	= 7x7	= 7^2
h	1 + 3 + 5 + 7 + 9 + 11 + 13 + 15	= 64	= 8x8	= 8^2
i	1 + 3 + ... + 17	= 81	= 9x9	= 9^2
j	1 + 3 + ... + 19	= 100	= 10x10	= 10^2
k	1 + 3 + ... + 21	= 121	= 11x11	= 11^2
l	1 + 3 + ... + 23	= 144	= 12x12	= 12^2

Fig 1a

Typical Jain Mathemagics Worksheet to discover the pattern formed
when consecutive odd numbers are progressively summed.

Notice the appearance of the Squared Number Sequence formed when
summing one more odd number to the last sum.

We can conclude that the progressive sum of odd numbers forms the sequence of squared numbers:

$$n^2 = 1, 4, 9, 16, 25, 36, 49, 64 \text{ etc}$$

Basically this data is suggesting that the expansion of numbers is systematically compressed into Form, or in this case, into Squared Shapes.

The concept being demonstrated here is that the **Supreme Language of Pattern Recognition** or **Universal Shape**, is operating.

The new mathematical education or revolution is the ability to see numbers as pictures or shapes.

It is the **Right Feminine Brain Mathematics** that understands pictures, shapes, music, holograms, patterns, symbols etc.

By learning this language, we are remembering the lost science of the Human–Dolphin Brain Connection. Research shows that Dolphins have the ability to do 8 things at once, like listening to 8 radio channels simultaneously!

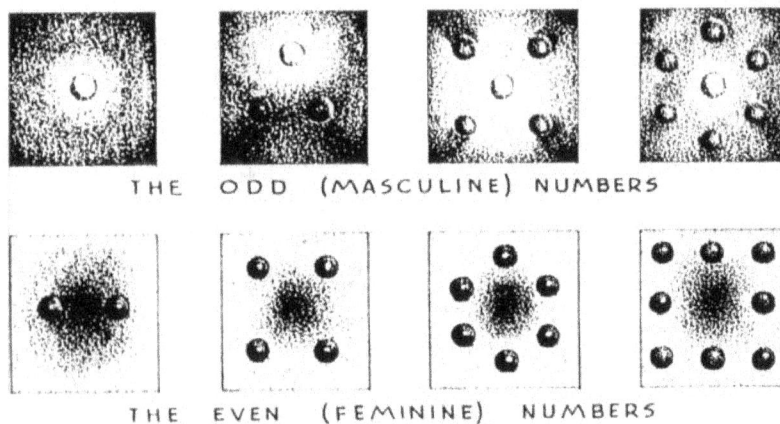

THE ODD (MASCULINE) NUMBERS

THE EVEN (FEMININE) NUMBERS

Fig 2

Pythagoras, from 2,500 years ago
attributed Male and Female qualities to the Series of
Odd (1-3-5-7-9) and Even (2-4-6-8) Numbers respectively.

PART 2:

ODD NUMBERS GENERATING
THE CUBED OR CUBIC NUMBERS

In Part 1, we saw how the Odd Nos. generated the Squared Nos.

To move **Into The Next Dimension**, (Jain's 17th **Sutra**) with this knowledge, is to know how to go from the Family of Squares into the Family of Cubes, and using only ODD NUMBERS !

Notice also how these sums are being added in a different way to the previous example. Instead of adding from the beginning number 1 at each row, we keep adding the next odd number in increasing amounts. The increase is always "By One More" so that the series naturally grows.

a	1	= 1	= 1 x 1 x 1	= 1^3
b	3 + 5	= 8	= 2 x 2 x 2	= 2^3
c	7 + 9 + 11			
d	13 + 15 + 17 + 19			
e	21 + 23 + 25 + 27 + 29			
f	31 + 33 + 35 + 37 + 39 + 41			
g	43 + 45 + 47 + 49 + 51+ 53 + 55			

Fig 3

Worksheet to discover the cubic display of odd numbers.

a	1	= 1	= 1x1x1	= 1^3
b	3 + 5	= 8	= 2x2x2	= 2^3
c	7 + 9 + 11	= 27	= 3x3x3	= 3^3
d	13 + 15 + 17 + 19	= 64	= 4x4x4	= 4^3
e	21 + 23 + 25 + 27 + 29	= 125	= 5x5x5	= 5^3
f	31 + 33 + 35 + 37 + 39 + 41	= 216	= 6x6x6	= 6^3
g	43 + 45 + 47 + 49 + 51+ 53 + 55	= 343	= 7x7x7	= 7^3

Fig 3a

Solution to the cubic display of odd numbers.

To develop a student's Mathematical Memory Power and Confidence, they are encouraged to learn these sequences, like the squared number sequence:

1 – 4 – 9 – 16 – 25 – 36 – 49 – 64 – 81 – 100 – etc

and the cubic number sequence:

1 – 8 – 27 – 64 – 125 – 216 – 343 – 512 – 729 – 1000 etc

Learning to view **Numbers as Shapes** is the first step to understanding the Galactic Mathemagics known here as THE ART OF NUMBER, or the Translation of Number Into Art.

Here is a final example, not about odd numbers, but showing another way how the Square Number Sequence can be generated.
Noticing the stair-casing of digits going up (ascending as in 1-2-3-4), then going down (descending as in 4-3-2-1).

This is a delightful example of pattern recognition.

1	=	$1 \times 1 = 1^2$
1 + 2 + 1	=	$2 \times 2 = 2^2$
1 + 2 + 3 + 2 + 1	=	$3 \times 3 = 3^2$
1 + 2 + 3 + 4 + 3 + 2 + 1	=	$4 \times 4 = 4^2$

Fig 4

Pyramid of Numbers

forming the Squared Numbers Sequence

CHAPTER 3

THE VISUAL MULTIPLICATION TABLE

Contents of this chapter:

- **PART 1**
  Digital Compression of the 4 Times Table (plugged into the Wheel of 9 to form the 9-Pointed Enneagram)

- **PART 2**
  "All The Ones" forming the Atomic Structure of Rutile Crystal.

- **PART 3**
  The superimposition of "All The Ones" with "All The Eights" forming the Atomic Structure of Platinum Crystal.

PART 1:
DIGITAL COMPRESSION
of the 4 TIMES TABLE

Let us convert the standard Multiplication Table (or Times Table) into surprising patterns of symmetry!

In this chapter we will explore the 4 Times Table to begin with, then apply a certain Unified Field to all the Digitally Compressed Digits of the Number "1" to arrive at Atomic Art!

To achieve this, we must first Digitally Compress or reduce all two digit numbers to single digits.

1	2	3	4	5	6	7	8	9
2	4	6	8	10	12	14	16	18
3	6	9	12	15	18	21	24	27
4	8	12	16	20	24	28	32	36
5	10	15	20	25	30	35	40	45
6	12	18	24	30	36	42	48	54
7	14	21	28	35	42	49	56	63
8	16	24	32	40	48	56	64	72
9	18	27	36	45	54	63	72	81

Fig 1

The well-known Multiplication Table up to 9x9.
Its inherent or hidden visual content is not taught in schools.

On the line below "Digitally Compress" the following two digit numbers of the 4 Times Table to single digits:

4   8   12   16   20   24   28   32   36   40   44   48   52   56   60   64   68   72

. . . .

Fig 2

Write down the digitally compressed form of the above 2-digit numbers that express the 4 Times Table.

What is the Periodicity of this Sequence? : . . . .

Fig 3

Write into the boxes above the digitally compressed numbers.
The 4 Times Table in its singular or Digitally Compressed Form shows a distinct recursive pattern or sequence of 9 Numbers that repeat forever.

To Transform Number Into Art, we need to **plug the repeating 9 numbers:**

4 - 8 - 3 - 7 - 2 - 6 - 1 - 5 - 9

into the **9 POINT CIRCLE,** or **Wheel of 9.**
In the space of Fig 4 below, draw a long unbroken line from 1 to 2 to 3 to 4 to 5 to 6 to 7 to 8 to 9 and by drawing a final connecting line from 9 to 1, as if to close the circuit.

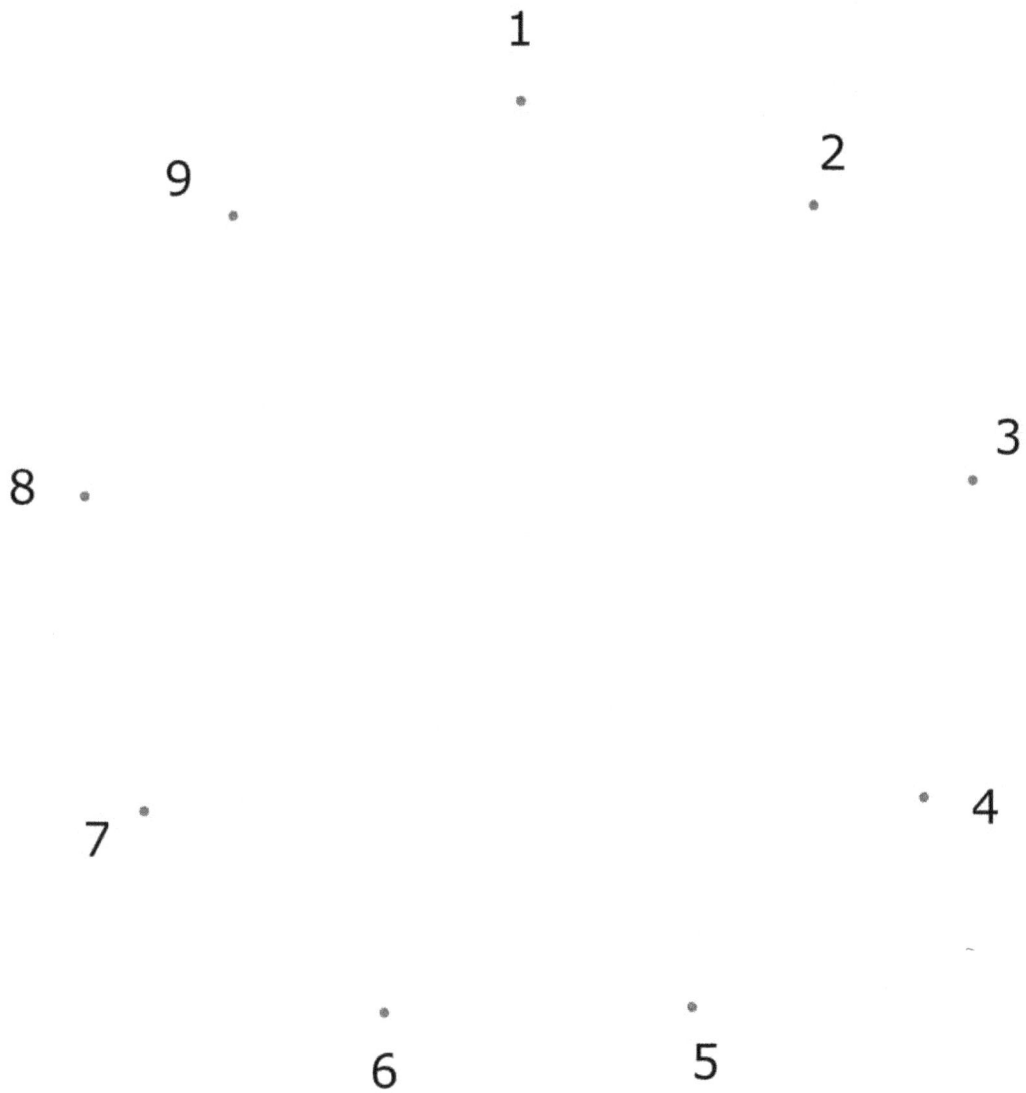

1

9

2

3

8

7

4

6

5

Fig 4

The Nine Point Circle is used to transform Number Into Art
To Convert the common 4 Times Table into a Picture.
The last number 9 is connected back to the first number 1.

59

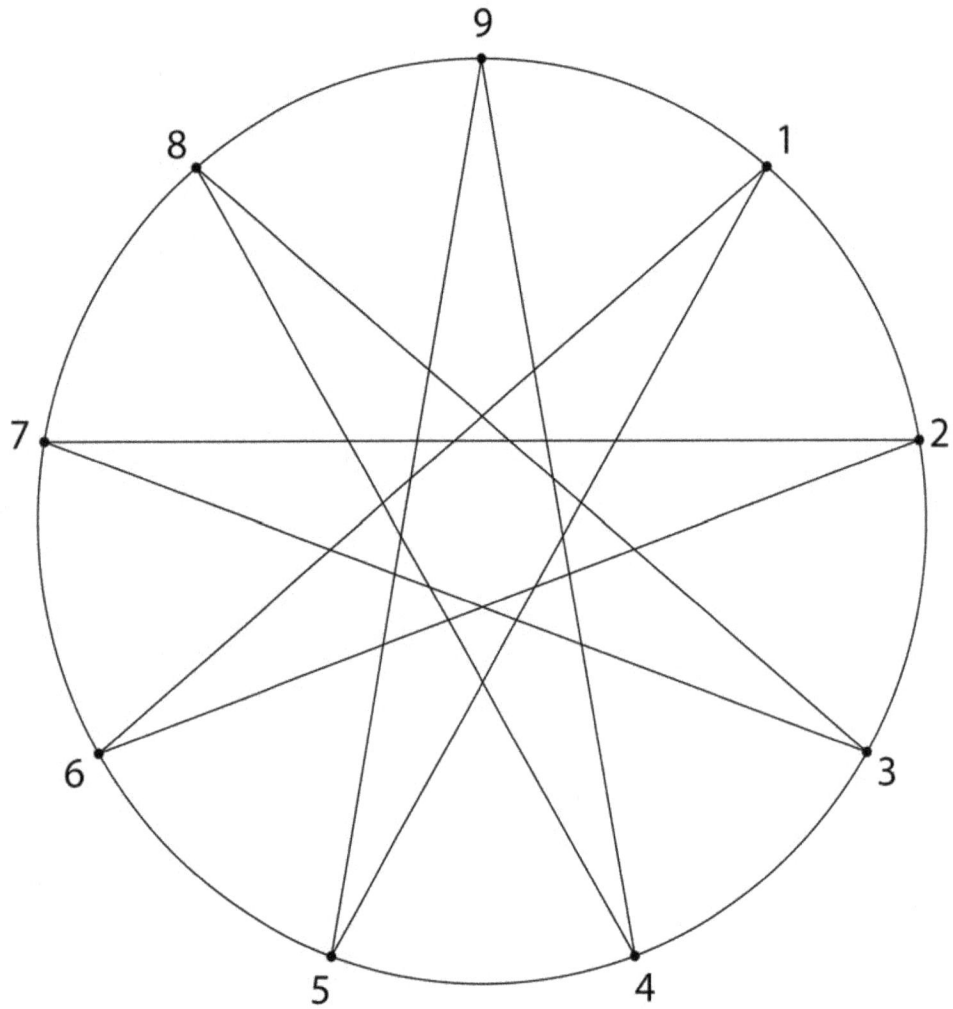

Fig 5

The Digitally Compressed 4 Times Table plugged into the Nine Point Circle generates the 9-Pointed ENNEAGRAM STAR

PART 2

"ALL THE ONES"
FORMING the ATOMIC STRUCTURE
OF RUTILE CRYSTAL

In Part 1, we just explored the Pattern of 4 which revealed a stellar symmetry.
In Part 2 we look at all the combined Patterns from 1 to 9, reduce them all to single digits, then isolate or highlight "All The Ones" and "All The Eights".
In Part 3, we superimpose or combine "All The Ones" with "All The Eights".

As part of the Jain Mathemagics Curriculum for the Global School, this ability to translate or turn or transform Numbers Into Art, that is, THE ART OF NUMBER, is an exciting part of Mathematics not taught in schools.
It is important that this Ancient Knowledge remains Universal, and that it is not referred to as The Vedic Square, that some Indian authors have attributed to it, it is far beyond the Vedas, it is Timeless and can not be blemished with Cultural Copyright.

Fig 1 showed the typical or traditional Multiplication Square up to 9x9.
We can now replace all 2 digit numbers with single digits.
eg: $6 \times 7 = 42 = 4 + 2 = 6$
Sometimes an extra step is required, as in:
$4 \times 7 = 28 = 2 + 8 = 10 = 1 + 0 = 1$

Here in Fig 6 is the space for you to create the Table of 81 Digitally Compressed Digits that births or generates a timeless matrix of ancient symbols. It is also referred to as **FIXED DESIGN**, as it can never change, it can belong to not one culture.

The word "Matrix" means "Womb" implying that this Box of Numbers, in its most primitive form, potentially gives births to highly intelligent sequences and derived images in the form of sacred symbols.

	1	2	3	4	5	6	7	8	9
2									
3									
4									
5									
6									
7									
8								1	
9									

Fig 6

**In this space,
complete the table to show the
Digitally Compressed Multiplication Table composed
essentially of 81 Single Digits from 1 to 9.**

An example has been done for you:
8x8 = 64
= 6+4 = 10
= 1+ 0
= 1

1	2	3	4	5	6	7	8	9
2	4	6	8	1	3	5	7	9
3	6	9	3	6	9	3	6	9
4	8	3	7	2	6	1	5	9
5	1	6	2	7	3	8	4	9
6	3	9	6	3	9	6	3	9
7	5	3	1	8	6	4	2	9
8	7	6	5	4	3	2	1	9
9	9	9	9	9	9	9	9	9

Fig 6a

**The Digitally Compressed Multiplication Table
composed essentially of 81 Single Digits from 1 to 9**

Fig 7a below is the space for you to highlight all the Number Ones contained in Fig 6a above.

In this regard, we are isolating "All The Ones", like thorough Pattern Hunters, not knowing what to expect, open to discovery in the hope of some symmetrical observation. Symmetry is the Language of the Universe!

Fig 8a below is the space for you to highlight all the Number Eights contained in Fig 6a above.

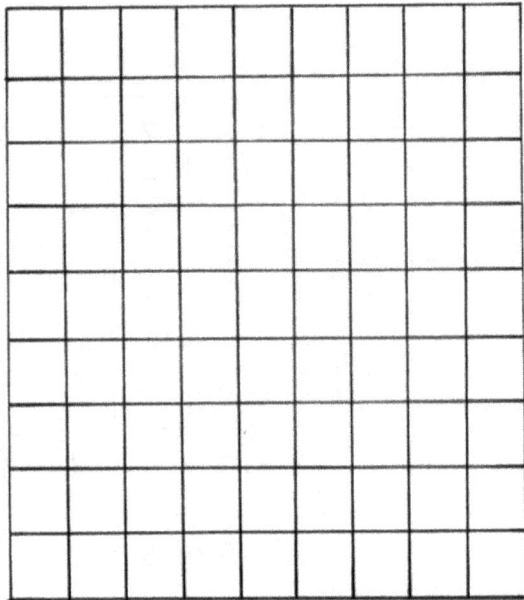

Fig 7a

Shade in, using any colour,
all the Digital Sums of "1"

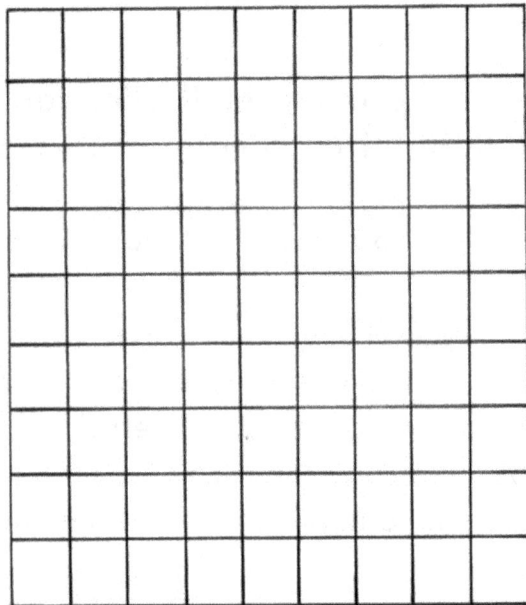

Fig 8a

Shade in, using a different colour than before,
all the Digital Sums of "8"

1	2	3	4	5	6	7	8	9
2	4	6	8	**1**	3	5	7	9
3	6	9	3	6	9	3	6	9
4	8	3	7	2	6	**1**	5	9
5	**1**	6	2	7	3	8	4	9
6	3	9	6	3	9	6	3	9
7	5	3	**1**	8	6	4	2	9
8	7	6	5	4	3	2	**1**	9
9	9	9	9	9	9	9	9	9

Fig 7b

Solution to All the Digital Sums of "1" which have been shaded in,
in the Digitally Compressed Multiplication Table.
For reference, we call it: "All The Ones".

At this stage, as we stare at the six highlighted ones, we can perceive or detect that there is a hint of symmetry, even though it may not look exciting. There is an important step or process coming that will extract a deeper meaning from these ones.

1	2	3	4	5	6	7	**8**	9
2	4	6	**8**	1	3	5	7	9
3	6	9	3	6	9	3	6	9
4	**8**	3	7	2	6	1	5	9
5	1	6	2	7	3	**8**	4	9
6	3	9	6	3	9	6	3	9
7	5	3	1	**8**	6	4	2	9
8	7	6	5	4	3	2	1	9
9	9	9	9	9	9	9	9	9

Fig 8b

Solution to All the Digital Sums of "8" which have been shaded in,
in the Digitally Compressed Multiplication Table.
For reference, we call it: "All The Eights".

If you can't see the connection or similarity between "All The Ones" and "All The Eights" lets try embellishing the Pattern for "All The Ones". To accomplish this we need to remove all the 81 numbers and all the 81 cells and just have a Grid of 81 Dots. You can visualize that every dot, shown in Fig 9 below, is really the centre of every numbered cell.

. . . .

. . . .

. . . .

. . . .

. . . .

. . . .

. . . .

. . . .

. . . .

Fig 9

The Grid of 81 Dots
that represents the 81 digitally compressed numbers
of the Multiplication Table.

We mark where those six Ones were located, say in a heavier dot, so that we can clearly see where their positions are, and wonder how they are all interconnected. When Einstein referred to his formulae in Relativity, he suggested that every atom in the universe is connected to every other atom, that there exists a Unified Field.

In this regard, every Point of One is inextricably connected to every other Point of One, so we need to draw a line or many lines that connects them all together.

All we are doing here is really Joining The Dots. (All I have done for the past 30 years really is Join The Dots! which has lead to this thesis called The Art Of Number).

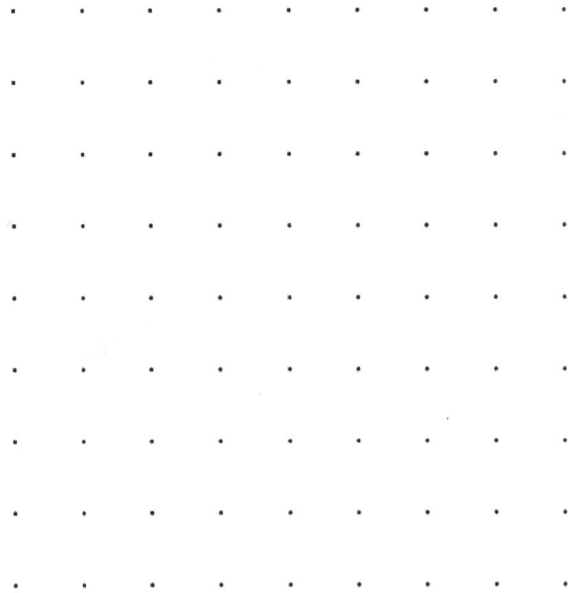

Fig 10a

In this space, connect "All The Ones" by Joining The Dots, such that every One is connected to every other One, to form a Unified Field of Oneness.

Similarly, lets look at All The Eights below in Fig 11b, so we can compare this visually to the pattern just generated above:

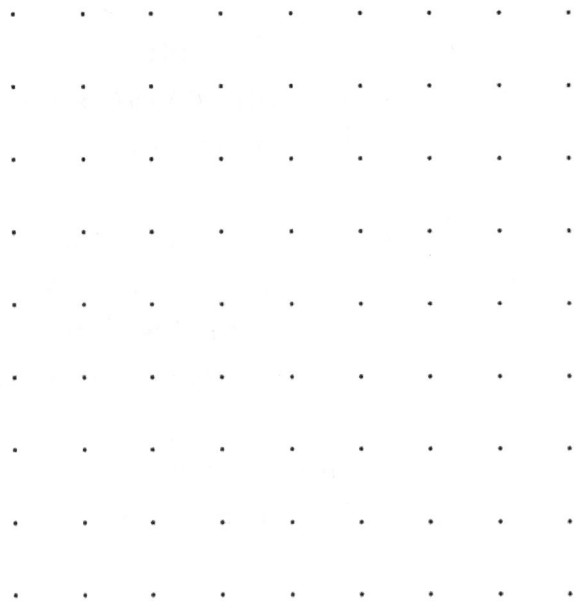

Fig 11a

In this space, connect "All The Eights" by Joining The Dots, such that every Eight is connected to every other Eight, to form a Unified Field of Eightness.

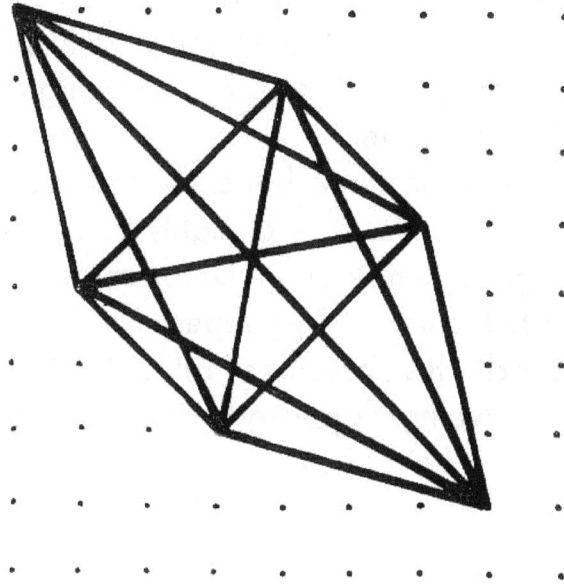

Fig 10b

Solution to the Pattern for "All The Ones" generated by Joining The Dots, such
that every One is connected to every other One,
to form a Unified Field of Oneness.

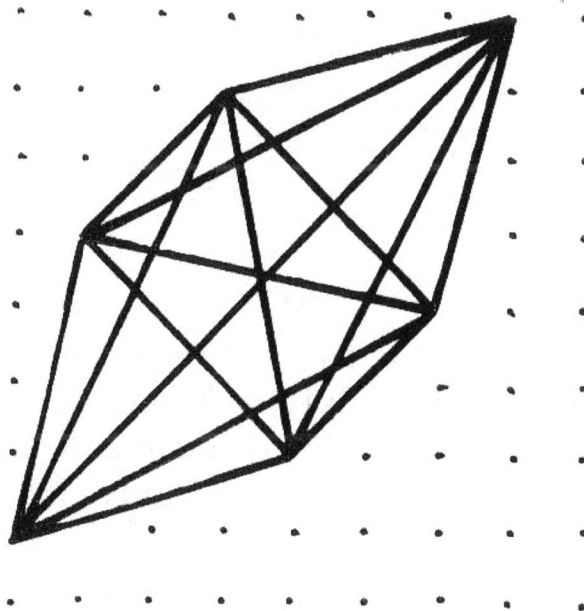

Fig 11b

Solution to the Pattern for "All The Eights" generated by Joining The Dots,
such that every Eight is connected to every other Eight.

When you research your old chemistry books, you will see that this Octahedral shape (the Octahedron is one of the 5 Platonic Solids, like a double square based pyramid joined base to base, and has 8 triangular faces) is the same shape as the crystal known as Rutile.

(Rutile is found in the sands of white beaches and is identified as the black streaks seen in these white sands. It is a highly sought after mineral used in satellites and other high technologies. Unfortunately, in the 1980s, big companies illegally started sand mining operations in Scott's Head in mid-coastal NSW, to extract this precious rutile, but were rallied against environmentalists who fervently opposed the destruction of the pristine beaches).

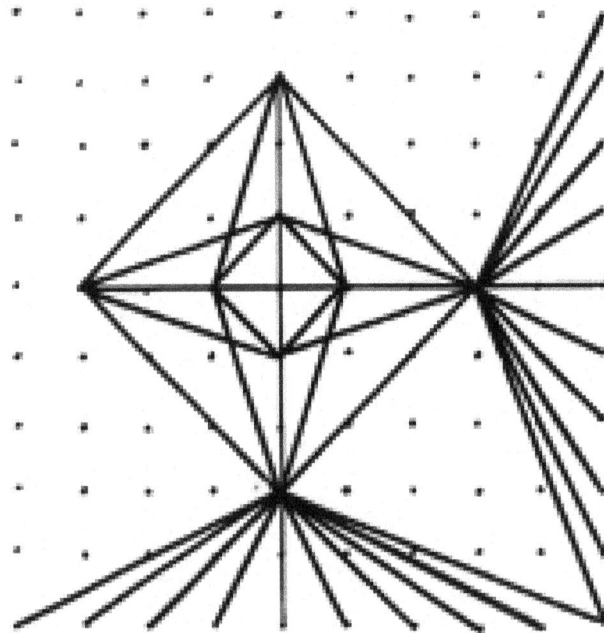

Fig 11c
The Pattern for "All The Zeroes"
seen in a typical 10x10 Multiplication Table,
exhibiting a typical crystalline octahedral shape similar to
the atomic structure of Rutile crystal.

PART 3
The SUPER-IMPOSITION of "ALL THE ONES" with the "ALL THE EIGHTS" FORMING the ATOMIC STRUCTURE of PLATINUM CRYSTAL

the VEDIC SQUARE

formerly called: "The Visual Multiplication Table and it's Reduction Grid".

1	2	3	4	5	6	7	8	9
2	4	6	8	10	12	14	16	18
3	6	9						
4	8							
5								
6								
7								
8								
9								

1	2	3	4	5	6	7	8	9
2	4	6	8	1	3	5	7	9
3	6	9	3	6	9	3	6	9
4	8	3	7	2	6	1	5	9
5	1	6	2	7	3	8	4	9
6	3	9	6	3	9	6	3	9
7	5	3	1	8	6	4	2	9
8	7	6	5	4	3	2	1	9
9	9	9	9	9	9	9	9	9

ALL THE ONES

ALL THE EIGHTS

Rutile

• DISCOVER HOW THE TRADITIONAL MULTIPLICATION - TABLE CREATES THE ATOMIC STRUCTURE OF:
1): RUTILE (on this page)
2): PLATINUM CRYSTAL (overleaf)
by understanding the Complements of 9 !
See "THE BOOK OF MAGIC SQUARES" Volume 3. Chapter "X".

Fig 12
This page shows a summary of what we have just understood,
isolating "All The 1s" and isolating "All The 8s".
The correct and Universal Name, is not "Vedic Square" but
"THE VISUAL MULTIPLICTION TABLE AND ITS REDUCTION GRID"
(title coined by Jain)

(image taken from one of my early books called
"The Book Of Magic Squares, Vol 3).

In this part, we will combine the two together, or superimpose the two images.

Step 1: Plot in all the Points of "1" in one colour, like red, and draw lines that connect each dot to every other dot representing "One-ness".

Step 2: Plot in all the Points of "8" in another colour, like blue, and draw lines that connect each dot to every other dot representing "Eight-ness".

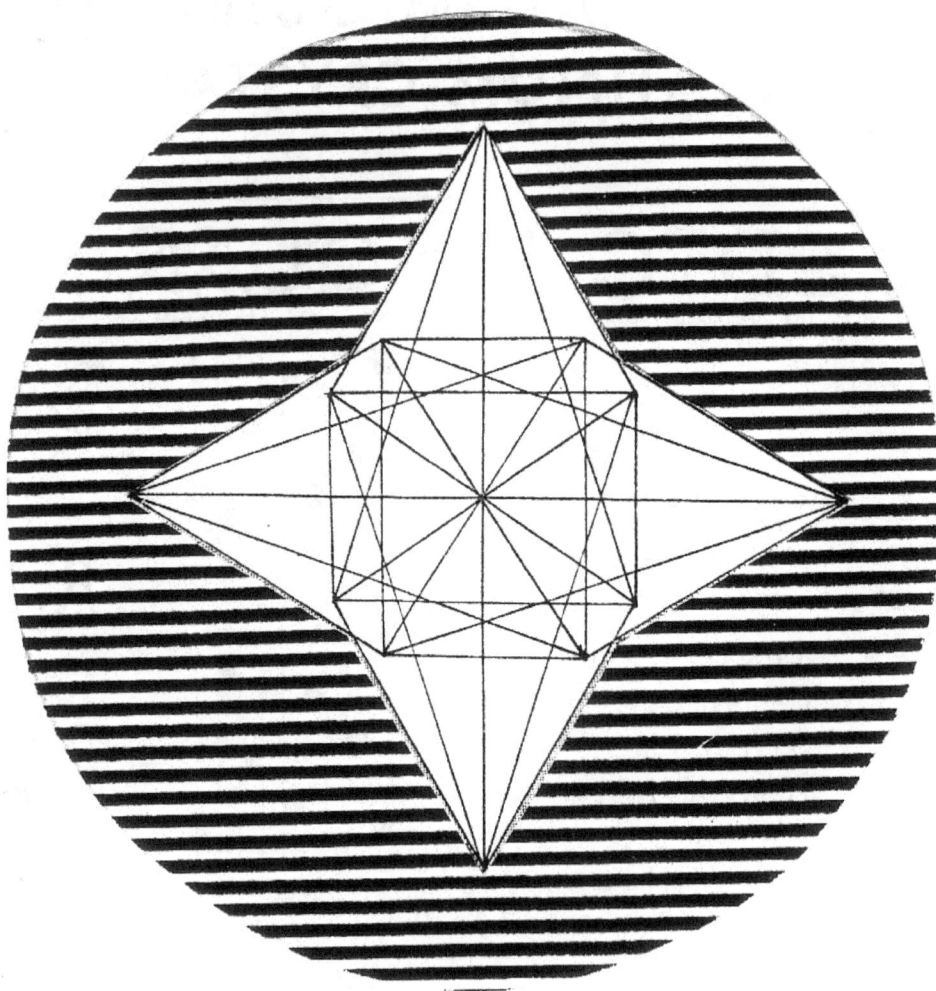

Fig 13
Superimposition of the "1"s and the "8"s.

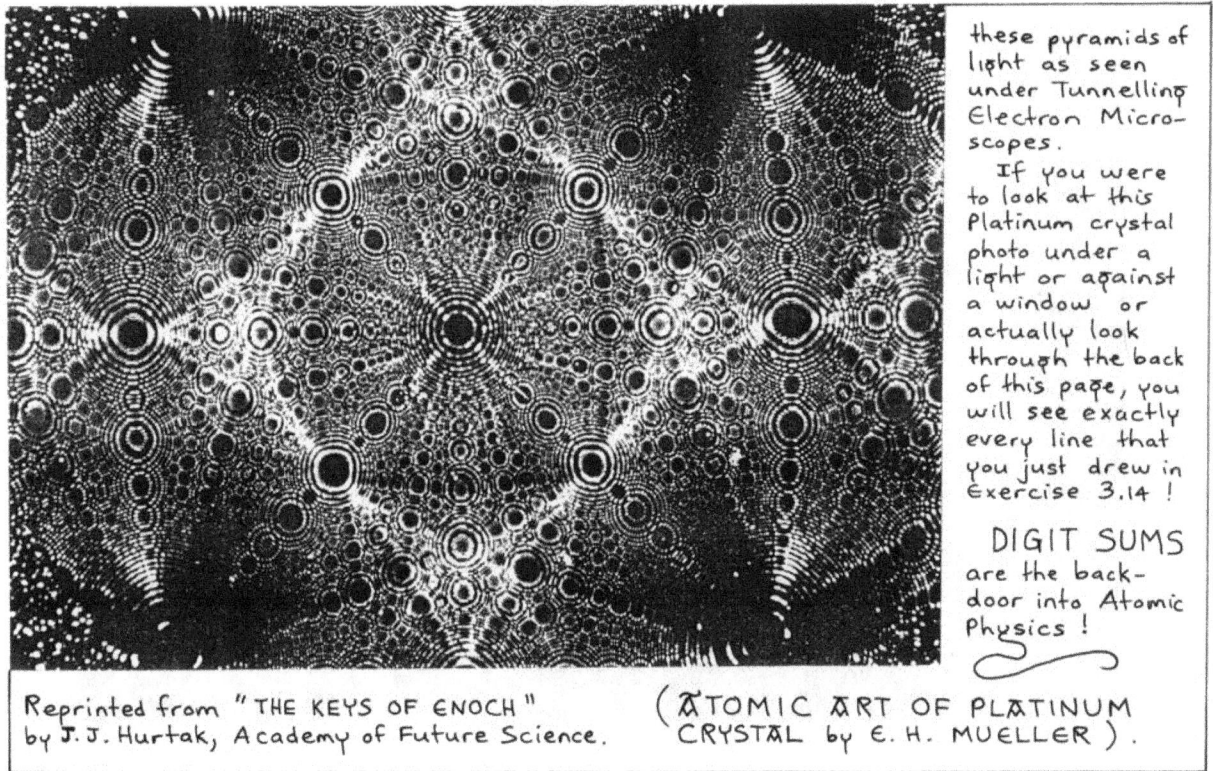

these pyramids of light as seen under Tunnelling Electron Microscopes.
   If you were to look at this Platinum crystal photo under a light or against a window or actually look through the back of this page, you will see exactly every line that you just drew in Exercise 3.14 !

DIGIT SUMS are the back-door into Atomic Physics !

Reprinted from "THE KEYS OF ENOCH" by J.J. Hurtak, Academy of Future Science.

(ATOMIC ART OF PLATINUM CRYSTAL by E.H. MUELLER).

Fig 14
The Atomic Structure of Platinum Crystal generated by the Superimposition of the "1"s and the "8"s of the digitally compressed Visual Multiplication Table.
(image reprinted from "The Keys of Enoch" by JJ Hurtak, Academy of Future Science).

The diagram or yantram or power art shown in Fig 13 is exactly this atomic art photograph. You can check this by looking though the back of this photograph and see clearly that every straight line is evident or visible and identical to the pattern of Fig 13.

This photograph slipped through locked internet sites.

It is the first time that humans have been able to successfully photograph the atom! The atoms can be seen as the little rings of concentric circles.

This photograph is resolved to almost capture the pyramid nature of Light.

Fig 15
In this very rare and historically important photograph
from the 1950s, Mueller is pointing to a section
of the Platinum crystal magnified 750,000 X.

Part of this research is to locate all the other atomic art photographs of all the possible Elements of the Periodic Table.
Students need to have access to all this Knowledge. What is the atomic structure of Gold? and Silver" and why are they not available in libraries?
Can we currently get access to all the works of Mueller, so that these amazing micro-photographs are shareable.

Fig 16
Murals of the combined Multiplications of "1"s + "8"s
painted by Jain in 1984 on 2m x 2m fabric sheets,
representing the Atomic Structure of Platinum Crystal.

All The Points of 1 + 8
8 x 22.5

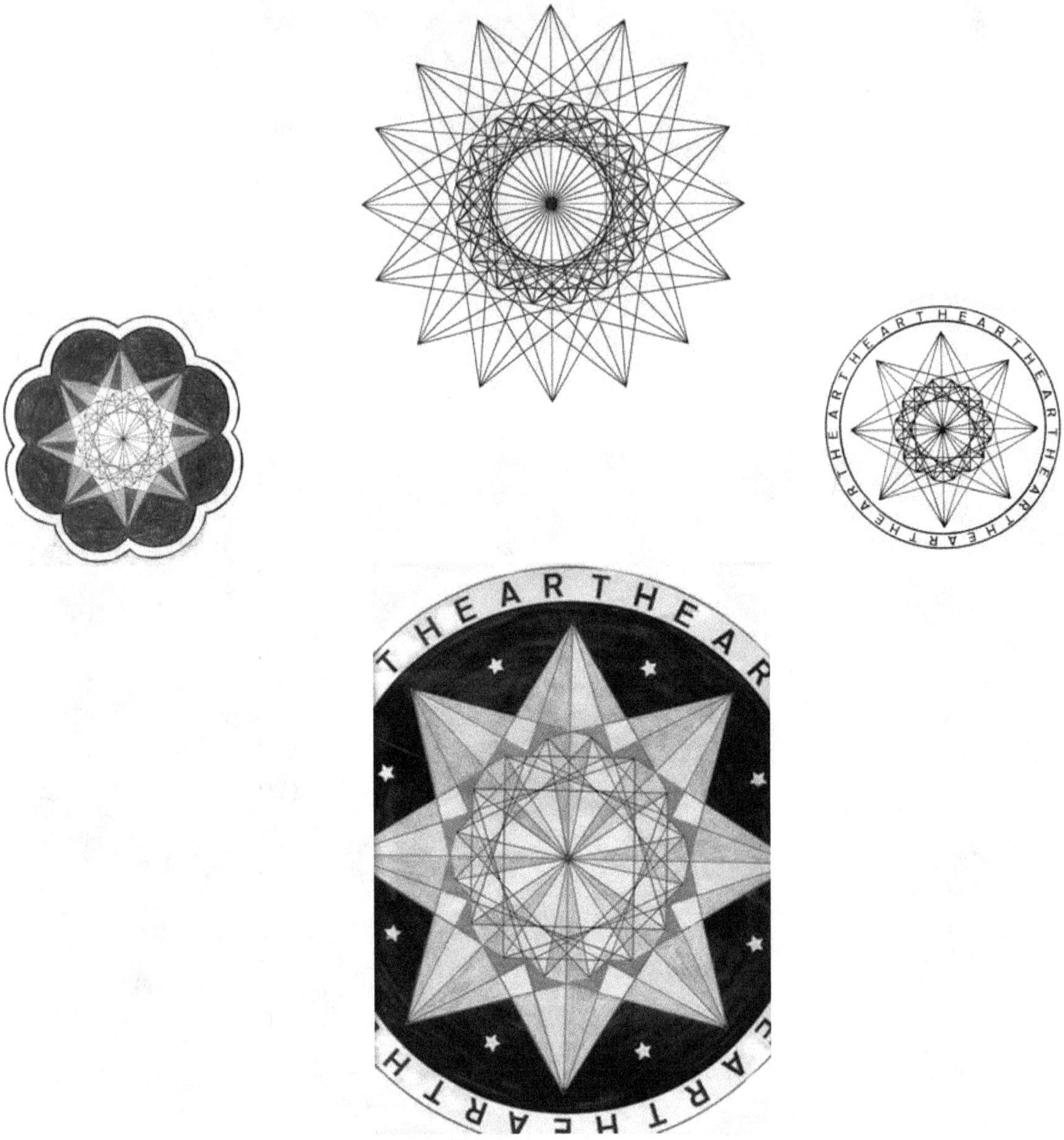

Fig 17
We can take this pattern of the digitally reduced Multiplication Table "1"s + "8"s and superimpose it again upon itself to arrive at a more complex or interesting crystalline pattern.
This is a future decal design available by 2013.

Decals project the sacred geometry into your house or onto your body, utilizing the magic of glass and sun to bring transformation, healing and re-alignment with the Divine Symmetries.

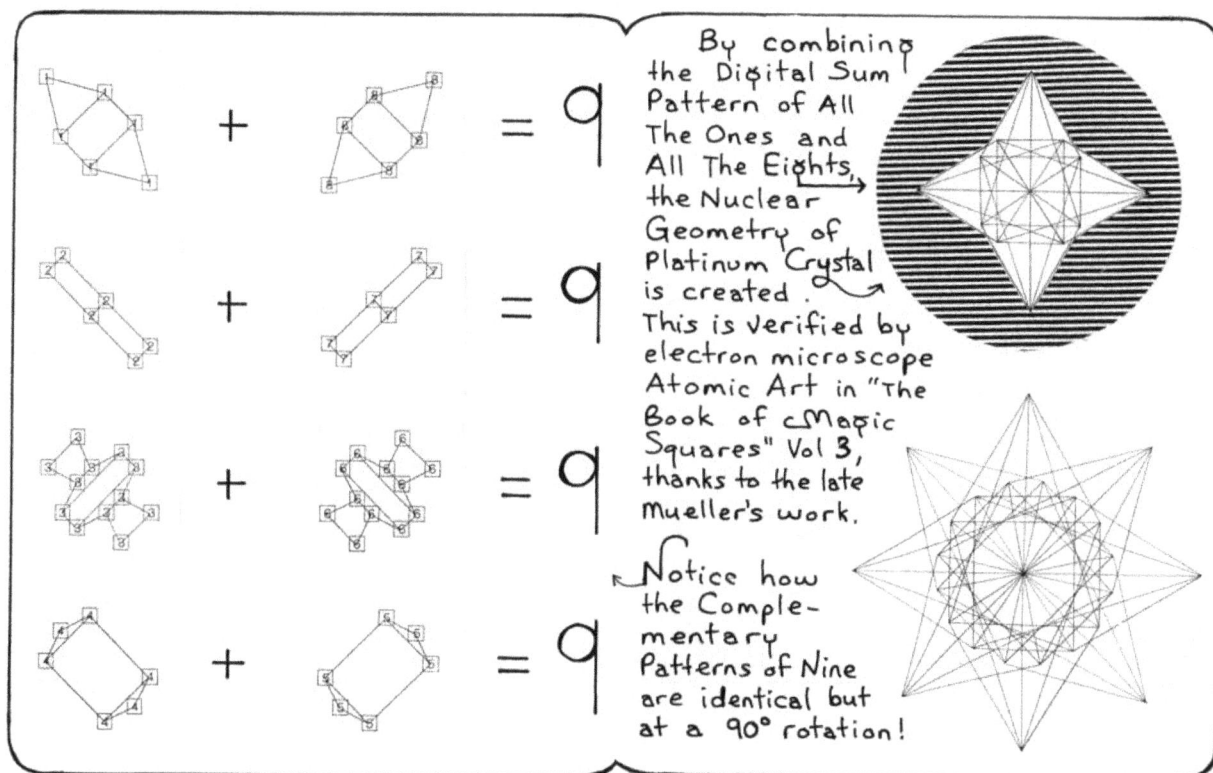

By combining the Digital Sum Pattern of All The Ones and All The Eights, the Nuclear Geometry of Platinum Crystal is created. This is verified by electron microscope Atomic Art in "The Book of Magic Squares" Vol 3, thanks to the late Mueller's work.

Notice how the Complementary Patterns of Nine are identical but at a 90° rotation!

Fig 18

Examination of all the other **Complementary Pairs of 9**.
Up till now, we have seen that the Ones and Eights
form a complementary Pair.
(This page is extracted from: The Book of Magic Squares, Volume 3, Jain)

In Fig 18 we see all the Pairs that have complementary patterns.

— The Pair of 1+8 are at a rotation of 90 to one another, as well as:
— The Pair of 2+7,
— The Pair of 3+6
— The Pair of 4+5

A another book can be offered that just deals with these 4 complementary pairs and showing all their creative applications and nuances.

The hand-drawn diagram on this page has been redrawn on computer on the following page to shown how this work has progressed over the decades of research.

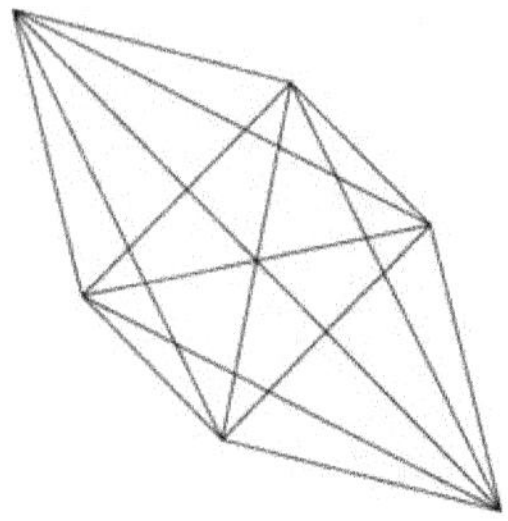	+	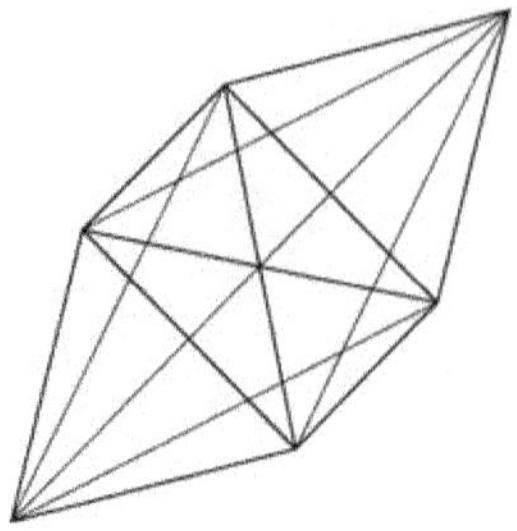	= 9
All The Ones		All The Eights	
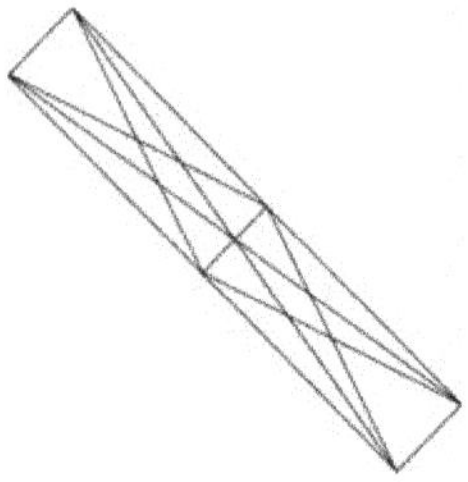	+	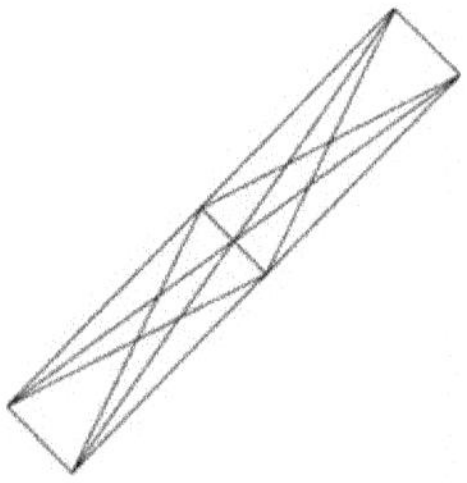	= 9
All The Twos		All The Sevens	
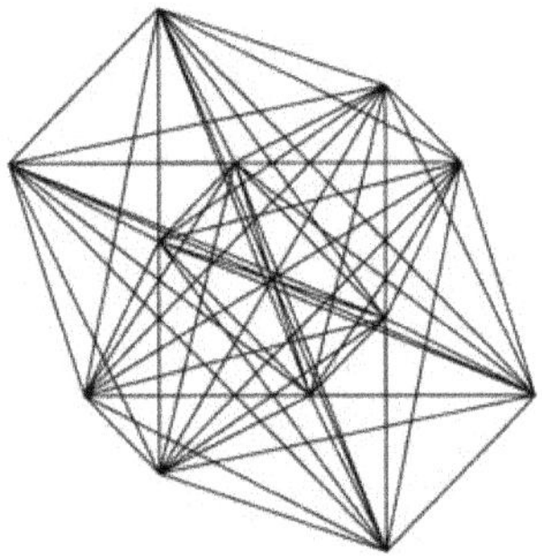	+	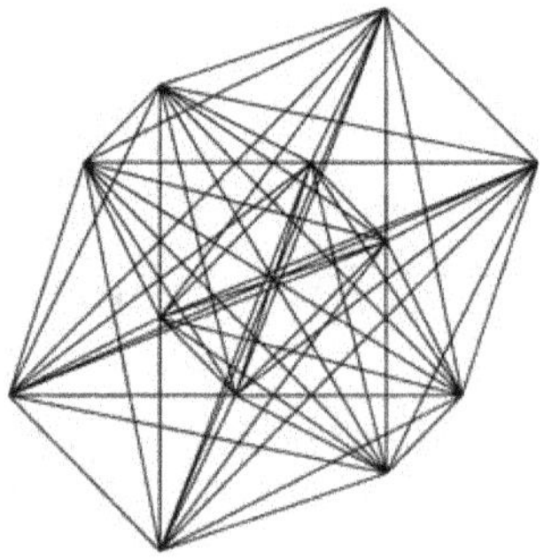	= 9
All The Threes		All The Sixes	
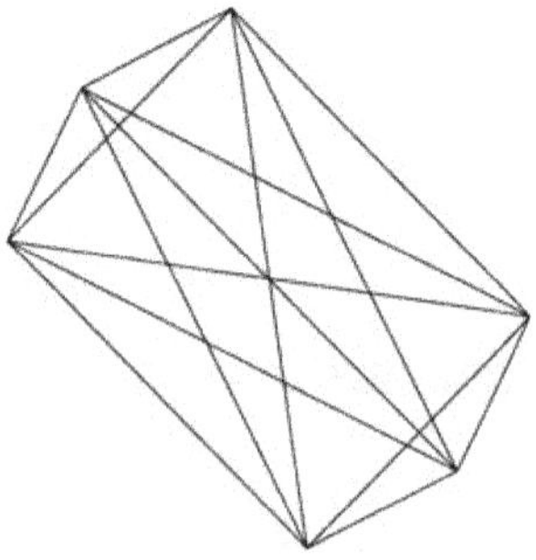	+	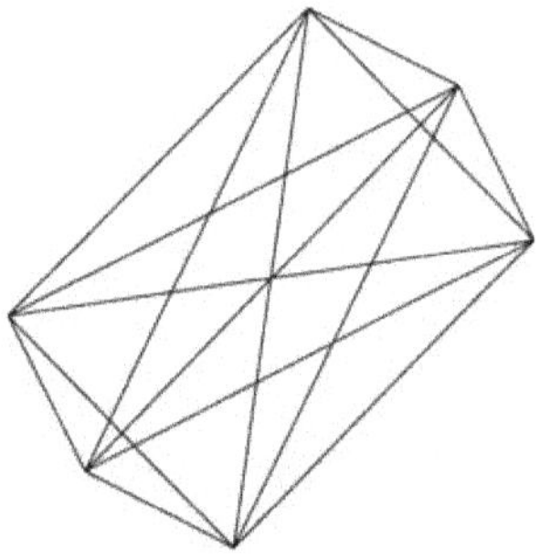	= 9
All The Fours		All The Fives	

Fig 18a
Chart of All The Pairs of the Visual Multiplication Table that sum to 9

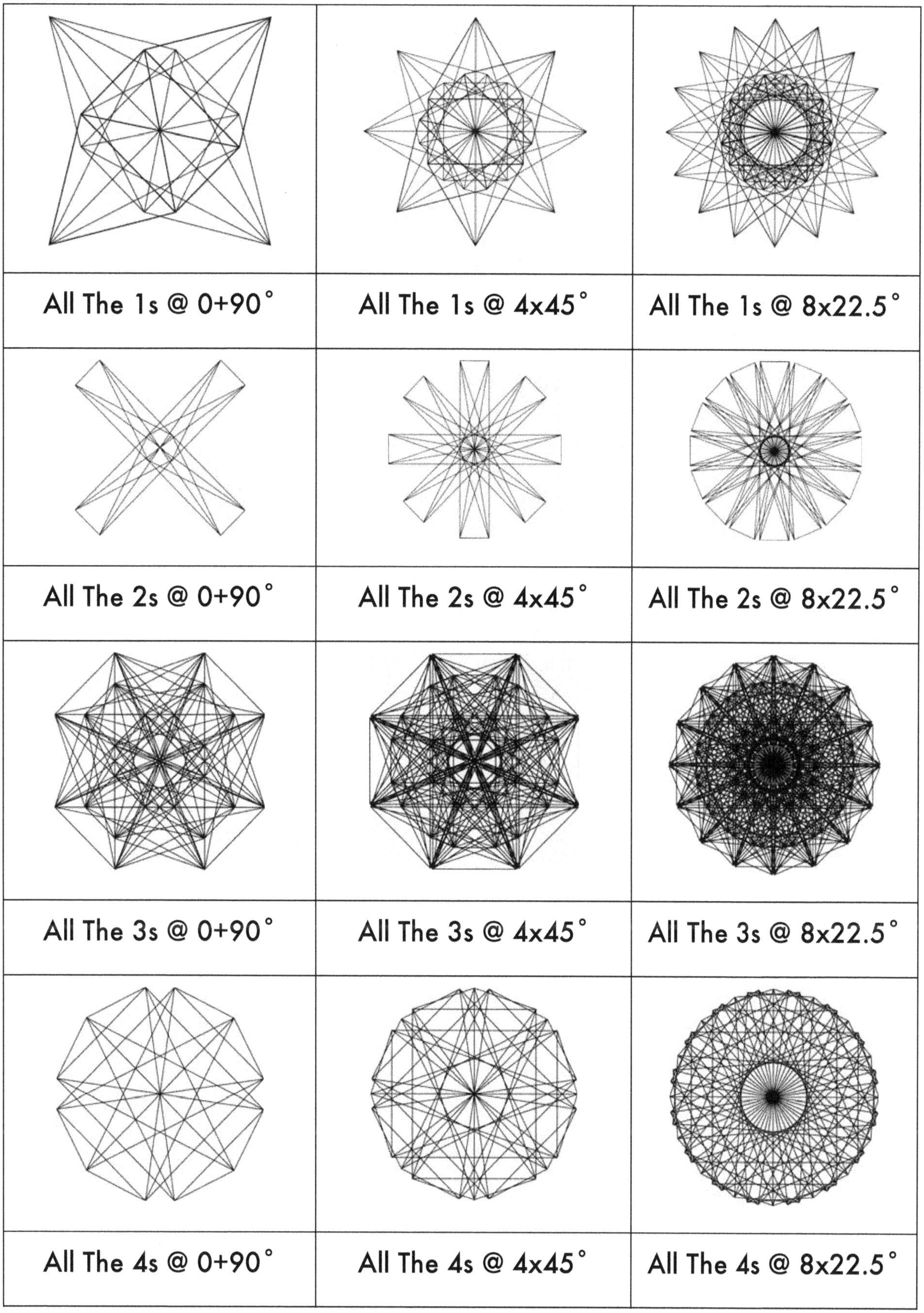

All The 1s @ 0+90°	All The 1s @ 4x45°	All The 1s @ 8x22.5°
All The 2s @ 0+90°	All The 2s @ 4x45°	All The 2s @ 8x22.5°
All The 3s @ 0+90°	All The 3s @ 4x45°	All The 3s @ 8x22.5°
All The 4s @ 0+90°	All The 4s @ 4x45°	All The 4s @ 8x22.5°

Fig 18b
Chart of All The Pairs of the Visual Multiplication Table Rotated
and Super-imposed at 2x, 4x and 8x.

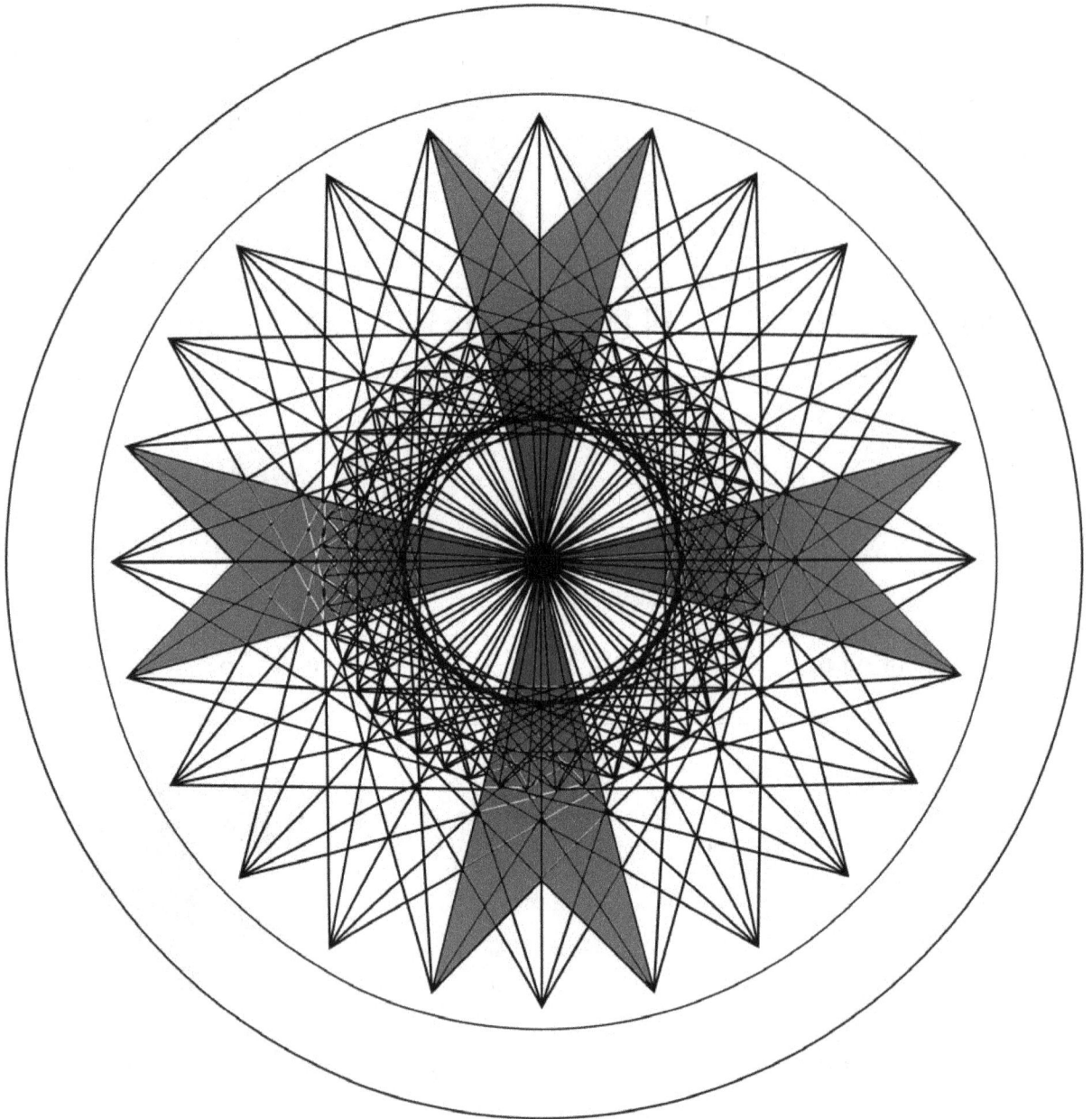

Fig 18c

All The 1s @ 12x15°

Prime Number Cross inspired from the previous chart in Fib 18b,
which shows All The Ones of the Visual Multiplication Table Rotated
and Super-imposed at 12x15 degrees
(because The Wheel of 24 when divided into the circle of 360 gives 15).
This is a very prized and highly psycho-active yantram (diagram or Power Art)
that can be used for future technologies.

Page 230

I whole-heartedly dedicate this Yantram of Fig 160 ("X" Grid of 1 • B at 0°+45°) to the AMBULANCE SERVICE of N.S.W. Australia 'cos they saved my Life on the dark nite of 27·1·1984 in the Blu Mountains, where I was stabbed close to the heart, 6" deep, whilst painting a 3m × 3m mural, similar to this design.
I have high-lighted the central cross to illuminate this sacred symbol of Rescue Healing.
G

Opening the   heart   chakra

fig 160

Fig 19

This page shows isolated patterns from the visual Multiplication Table.
Most of this material is extracted from my handwritten book:
THE BOOK OF MAGIC SQUARES, volume 3.

Page 180

he KNIGHTS TEMPLARS wore the Cross pointing to the four quarters in token of the Power by which a man may rise to loftier consciousness & which is found at the root of every form in the Universe.

Illustrated here is the Cross & the Cipher → ("Cipher" = a secret method of writing as by a specially formed code of symbols). The Cipher writing is reputedly derived from the component parts of the eight-pointed Cross. It is really a most ingenious device & yet simple, the key to which each Knight wore upon his breast. This is shown in Fig 159. The first four diagrams show how the Cipher was built up. The fifth "V" shows that "passing thru the Octagon" was "standing on the Cross".

The 8-pointed Cross is illustrated in this Chapter "X" 'cos this shape appears commonly: (Fig 153 at 0°+45° shown here on the top Right-hand side of this page = Fig 160). This 'Maltese Cross' also appears prevalently in other majik Square Art centerpieces.

"THE BEST WAY TO LEARN ANYTHING IS TO DISCOVER IT BY YOURSELF"

Fig 159

Shown above is Fig 160 which you will draw next. Known also as The "X" of 1 (Fig 151) drawn 4 Times at 45° angles Yantram.

Around the banner I have painted the word "EARTH" in a necklace 14 times, so that now it reads "HEART HEART HEART" etc. It can also read: "HEAR THE ART" etc. This was originally painted on a banner 11 years ago. ©

81

On one hand, Vedic Mathematics or Rapid Mental Calculation can be given a green tick, it works well in **Base 10**, making calculations very easy, so we can say that EARTH MATHEMATICS IS BASE 10.

On the other hand, we can say **BASE 9**, as observed in these last exercises, is a **GALACTIC MATHEMATICS**, along with its partner 12, and can conclude that: **BASES 9 and 12 are GALACTIC BASES.** (12x9=108)

Part of the vision of our Temple of Mathematics (EarthHeart) is to ensure that all children have the opportunity to learn equally as much about Base 9 and Base 12 as they do with Base 10 at conventional schools.

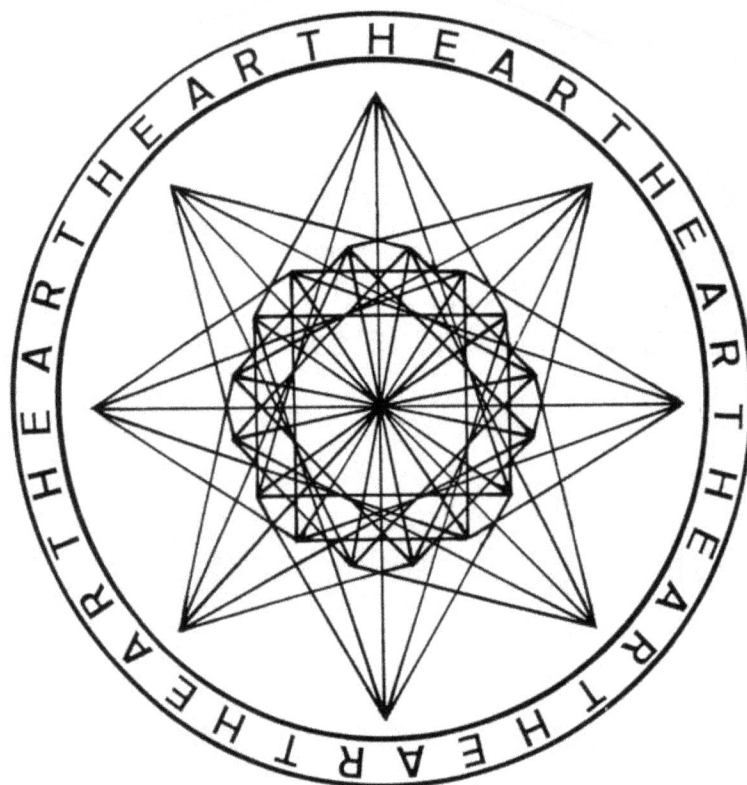

Fig 20

"ALL THE POINTS OF ONE" ROTATED 4 TIMES AT 45°"
You can colour in your own pattern here that stands out for you.
You are looking for two things: shape and colour.
Often after colouring in a Soul Yantram like this,
it is really a reflection of your auric field.

"ALL THE POINTS OF NINE"

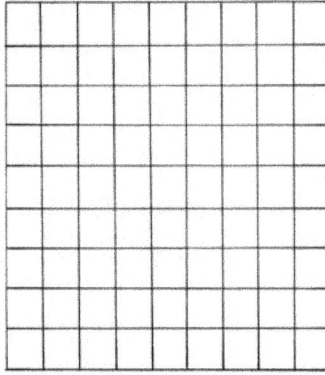

Fig 21

In the space above, write in all the number 9s, or just shade in the cells that represent "9" in the digitally compressed Multiplication Table shown in Fig 6a.

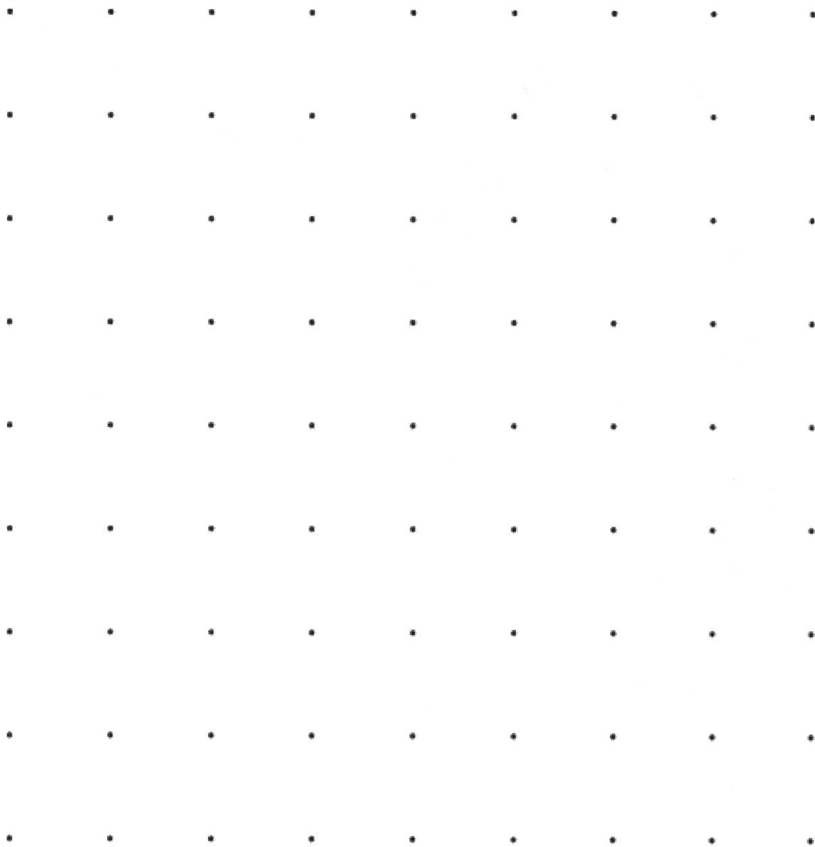

Fig 21a

Upon the Grid of 81 Dots draw the pattern for "All the 9s" ensuring that each dot is connected to every other dot.

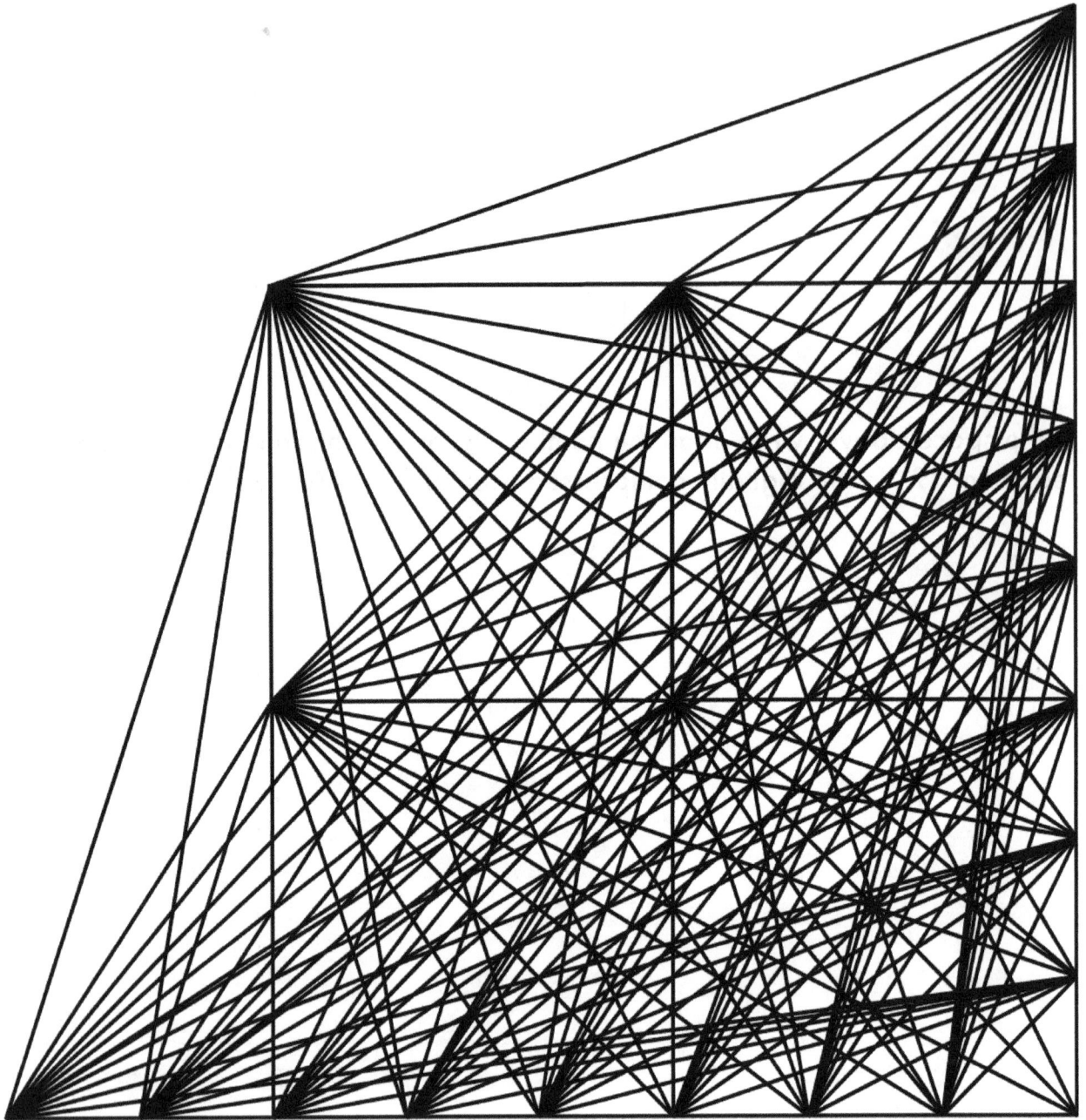

Fig 21b
"All the "9s" from the Visual Multiplication Table,
where each point of 9 is systematically connected
to every other point of 9 to create this Unified Field of 9ness.
(computer generated)

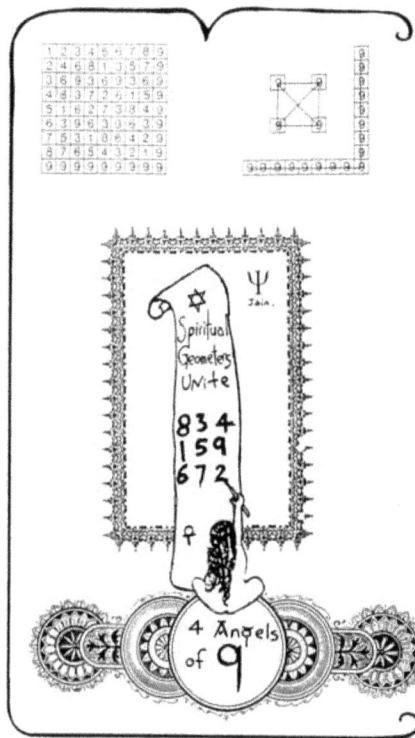

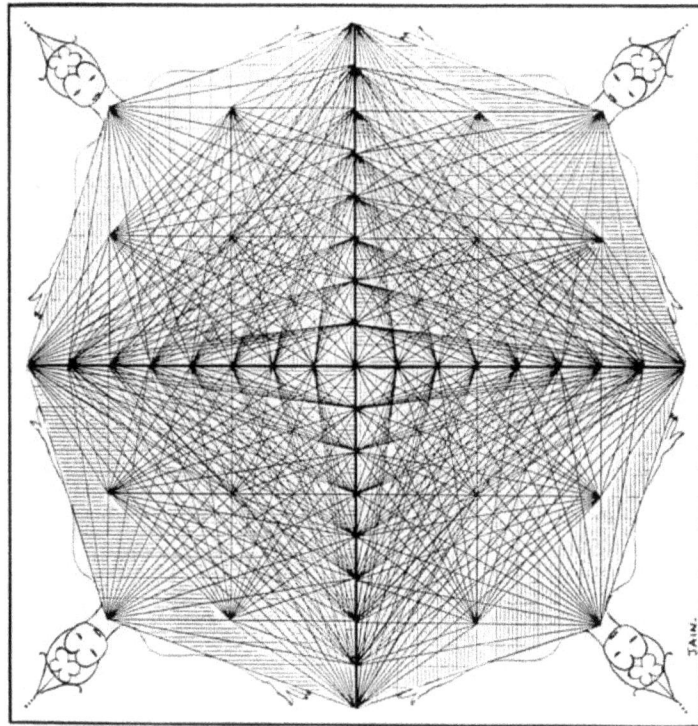

Fig 21c
"THE FOUR ANGELS OF 9"
(hand-drawn by Jain 108 circa 1985)
Here we have "All the "9s"
tessellated or tiled 4 times at various 90 rotations
to make a creative expression symbolizing the Earth's 4 Angels.

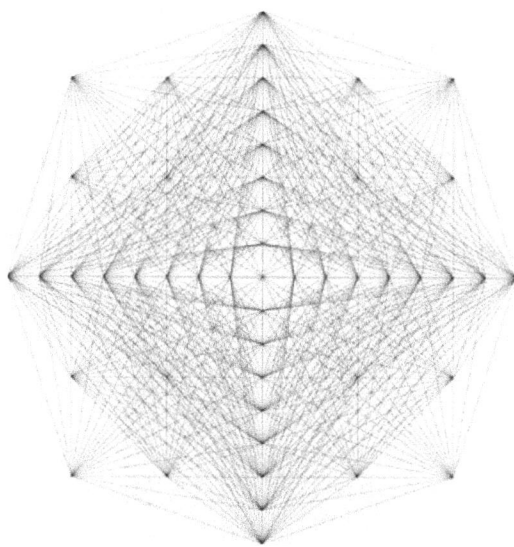

Fig 22

COLOUR TILING OF THE MULTIPLICATION TABLE

Select 9 various coloured pencils of your choice and appropriate one colour to each of the numbers from 1 to 9.

eg: the colour Red could be "All the Ones" in the digitally compressed Multiplication Table of Fig 6a.

eg: the colour Orange could be "All the Twos" in the digitally compressed Multiplication Table of Fig 6a. etc.

Observe the black "L-Shape" for "All The 9s" below:

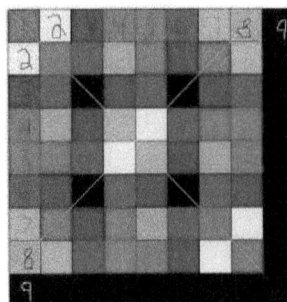

Chapter 4

THE FIBONACCI NUMBERS GENERATING THE PHI SPIRAL

+ LOST SECRETS OF THE 108 PHI CODE

PART 1
WHAT YOU ALREADY KNOW
ABOUT PHI and The FIBONACCI SEQUENCE

PART 2
WHAT YOU DON'T KNOW
ABOUT PHI and The FIBONACCI SEQUENCE:
JAIN'S DISCOVERY of The REPEATING 24 PATTERN

PART 3
WHAT YOU NEED To KNOW
ABOUT PHI and The FIBONACCI SEQUENCE
TIME CODE 12:24:60 ENCRYPTED
in The FIBONACCI SEQUENCE and PASCAL'S TRIANGLE

PART 4
The BINARY CODE Vs The PHI CODE

PART 1

WHAT YOU ALREADY KNOW
ABOUT PHI and The FIBONACCI SEQUENCE

Most people are now aware of the importance of the Fibonacci Sequence, thanks to the best selling book: "The Da Vinci Code", by Dan Brown, whether or not the historical facts are true or false is not as important as how far-reaching the effect this book has had on the mass-consciousness. The success of this book means that a few more million people have heard of the once well-known Fibonacci Sequence of numbers:

0, 1, 1, 2, 3, 5, 8, 13, 21, 34, 55, 89, 144 etc

They are derived by having a starting point of 0 and 1, the substance and binary language of all computers, and adding these two beginning numbers together:

0 + 1 = 1

This gives the third number of the sequence: 0, 1, **1,**

Then add the previous "1" to the last "1" which gives: 1 + 1 = 2

Giving the fourth number of the sequence: 0, 1, 1, **2,**

It's like adding the Past, to the Present, to give the Future, a veritable Trinity of Numbers, thus the next number would be:

1 + 2 = 3 and the continuing sequence becomes 0, 1, 1, 2, **3,** and so on.

Our task is to convert these Fibonacci Numbers into an Artform.

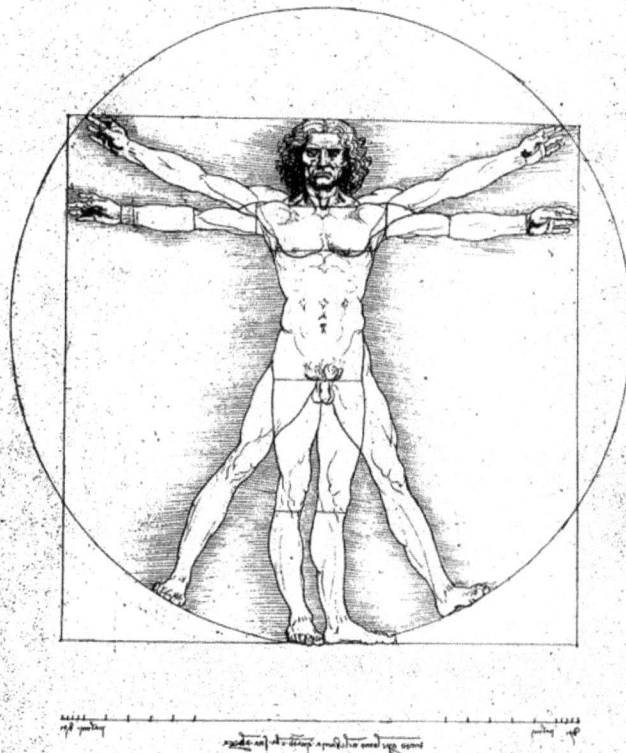

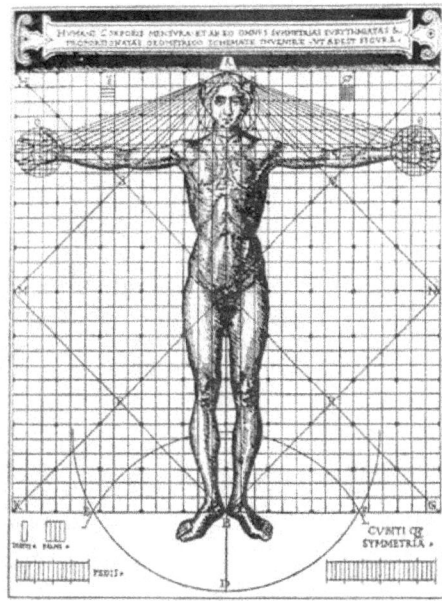

Fig 1

**The Human Body is the ultimate expression of the Phi Ratio.
Where the elbow bends,
compared to the whole length of the human arm,
is in the Divine Proportion, expressed mathematically
as 21:34 of the Fibonacci Sequence.**

(Shown here on right, is the Vitruvian Man, (taken from Manly P Hall's classic book:
"Secret Teachings Of All Ages").

When we decimalize these ratios, they approach the number: 1.618033988... and travels to infinity. We call this relationship, when indexed against Unity, as the Phi Ratio: **1:1.618033988...** and symbolize it by using a Greek letter of the alphabet that gives the "f" sound, called "**Phi**" or "φ" (whereas most people are familiar with the "p" sound in the Greek language, called "Pi" or "π"). Some people incorrectly pronounce "Phi" as "fee" but I write it here to get it right for the future generations of Fibonatics that the correct pronunciation is expressed as the same sound heard in the nursery rhyme: "Fee, **PHI**, Fo, Fum, I smell the blood of an Englishmun".

The Phi Ratio is the living mathematics of Nature.

We see these Fibonacci Numbers in the pentacle form of many flowers like the passionfruit; we see it in the Pine Cone, 8 spirals going one way intercepted by 13 spirals going the other way, and in the Sunflower, we see 21 Clockwise spirals versus 34 anti-Clockwise spirals.

89

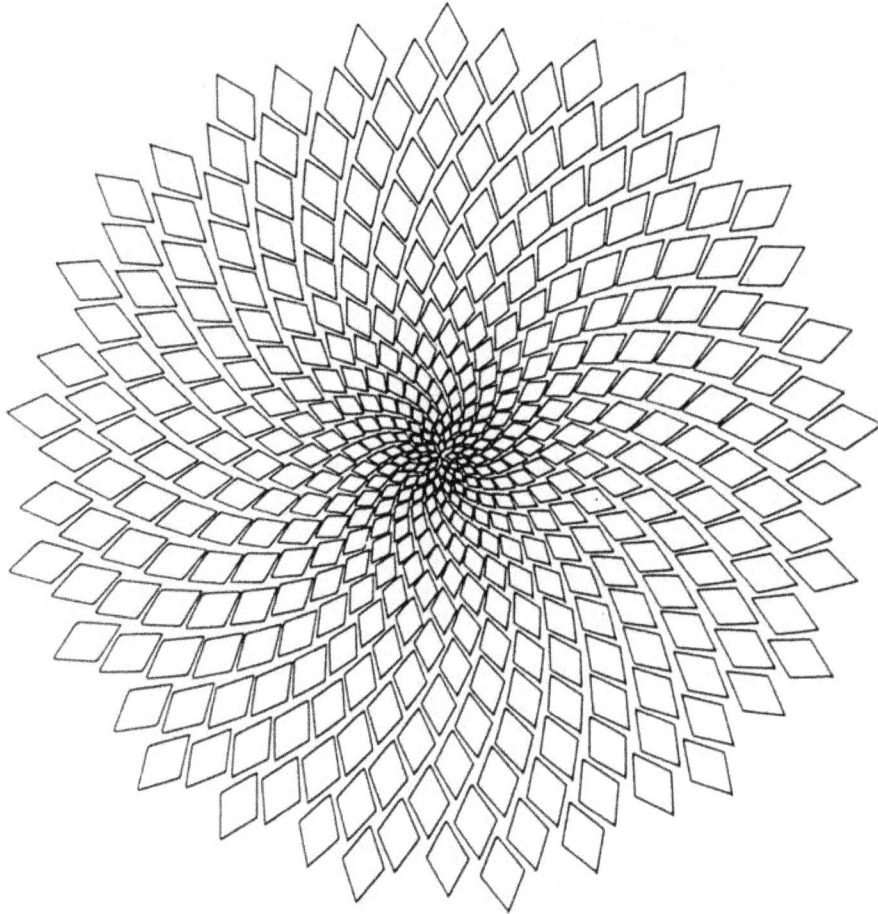

Fig 2

**Computer Rendition of the Sunflower Floret
showing 2 distinct spirals:
34 Anti-Clockwise and 21Clockwise.
This is the most PsychoActive diagram that I know,
from 30 years of research, this image depicting counter-rotating
fields, is the Law of Nature, and the Universe.**
(this image is a beautiful computer-rendition taken from a book by Robert
Dixon, p364 image "sunflower disc phyllotaxis with sigmoidal growth).

We also see these numbers as the approximate distances of the planets from the
sun, known in the scientific world as Bode's Law. Why are we awe-struck when
we view **Sacred Architecture** like the Parthenon of Greece and the Pyramids
of Gizeh. It has to do with Shape: the façade of the Parthenon is a Golden
Rectangle, in the proportion of 21:34, which is identical to the proportion of the
Human Body like where the elbow bends, where the knee bends, thus temples
are reflecting our own internal symmetry, so we are literally seeing ourselves
when we view beautiful buildings constructed in this awareness.

So for thousands of years, great artists used this Fibonacci Sequence as the blueprint for their canvass. That is why we are attracted to the proportions of the famous **Mona Lisa,** as the outer rectangle that contains her form has the blueprint of the Golden or Phi Rectangle in the ratio of 21:34. In fact, she has become the global embodiment of the Phi Ratio.

When I was in Tibet in 1990 I was with an artist who was painting his Thangkas or holy images of Buddha, and noticed the book that was the basis of all his measurements. It was referenced with the Fibonacci Numbers, making it law that every Buddha must be drawn in the Phi Rectangle otherwise it was considered bad energy. The classical book on this subject that introduces you to the universality of the Golden Mean as found in Nature is:
"The Power of Limits" by Gyorgy Doczi, subtitled: Proportional Harmonies in Nature, Art and Architecture, an example of the Buddhist's canon is shown below.

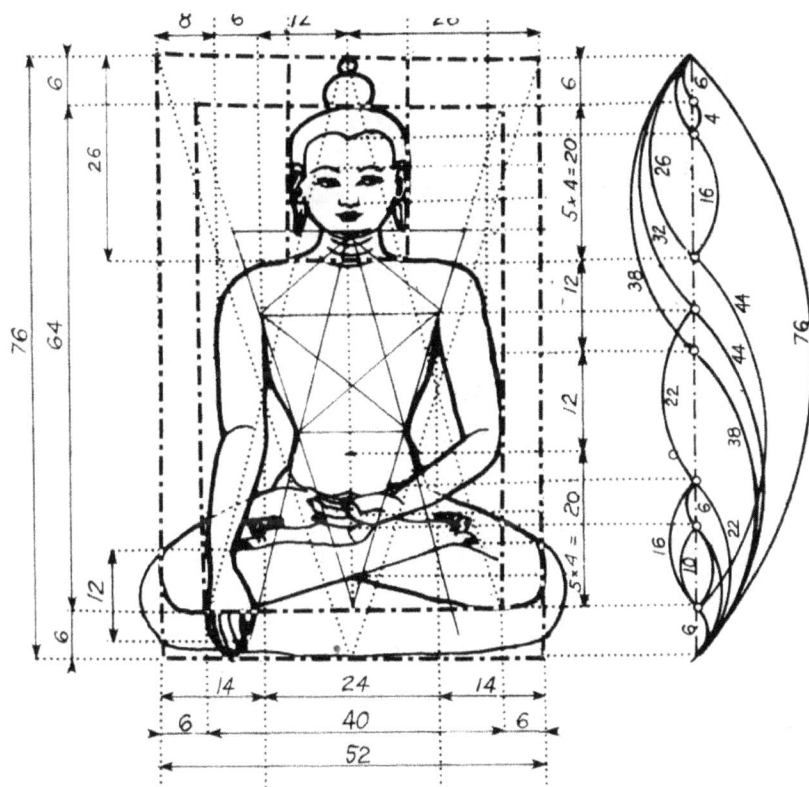

Fig 3

3 Golden Rectangles in the Biometrics on Buddha's Proportions.

Here is a very powerful or psycho-active diagram that changes consciousness, a diagram that brings together the worldview of the Scientist (factual measurements of the helical DNA) and the Metaphysician (someone who intuitively understands the higher meaning of the Phi Ratio). This bridging of 2 worlds is part of the change going on the planet now. When we view the twisting DNA molecule as it fits inside of a cylinder, and measure the critical distance from one atom to align itself to its original position after one complete turn or rotation around the central axis, like a spiraling vine would, our dear scientists noticed that the diameter was exactly 20 AU (Angstrom Units = one ten-millionth of a metre) and the vertical axial distance was 34 AU. You will notice that this relationship of 20:34 is almost identical to the Phi Ratio of 21:34. Metaphysicians are intrigued by this DNA Phyllotaxis and the mathematics of DNA approximating the Phi Ratio. They conjecture that perhaps we are not currently in the Phi Ratio, but are evolving to that ideal state. (A more paranoid interpretation would be that Someone, or some Race, in our distant past, created us humans as slaves in goldmines, engineered by the Extra-Terrestrial Masters of Gene-Splicing who were harvesting our planet for our DNA, and had genetically engineered as thus). This is one purpose of Sacred Geometry, is to start asking questions that bridge beliefs, knowledge and history of who we are.

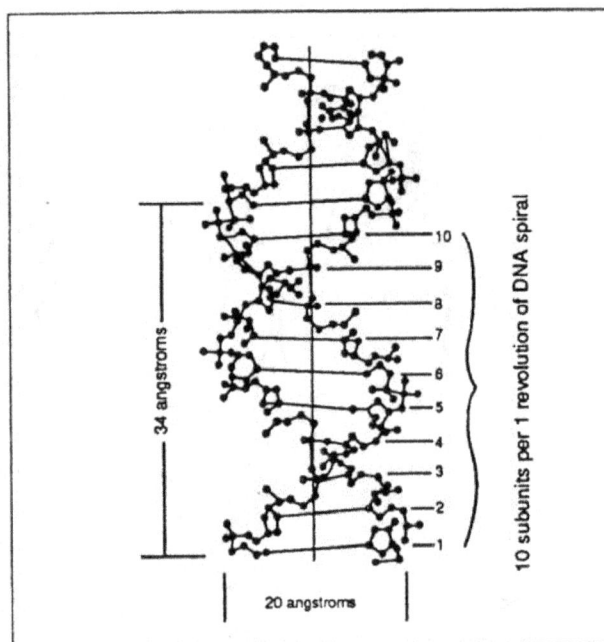

One 360 degree turn of DNA measures 34 angstroms in the direction of the axis. The width of the molecule is 20 angstroms, to the nearest angstrom. These lengths, 34:20, are in the ratio of the golden mean, within the limits of the accuracy of the measurements. Each DNA strand contains periodically recurring phosphate and sugar subunits. There are 10 such phosphate-sugar groups in each full 360 degree revolution of the DNA spiral. Thus the amount of rotation of each of these subunits around the DNA cylinder is 360 degrees divided by 10, or 36 degrees. This is exactly half the pentagon rotation, showing a close relation of the DNA sub-unit to the golden mean.

Fig 4
DNA Molecule, as it turns helically, exhibits approximately the 21:34 Phi Ratio

When we convert these Fibonacci Numbers into 2-Dimensional squares, they fit snuggly into the Golden Rectangle whose subsequent quarter-circle arcs form the familiar Golden Mean Spiral:

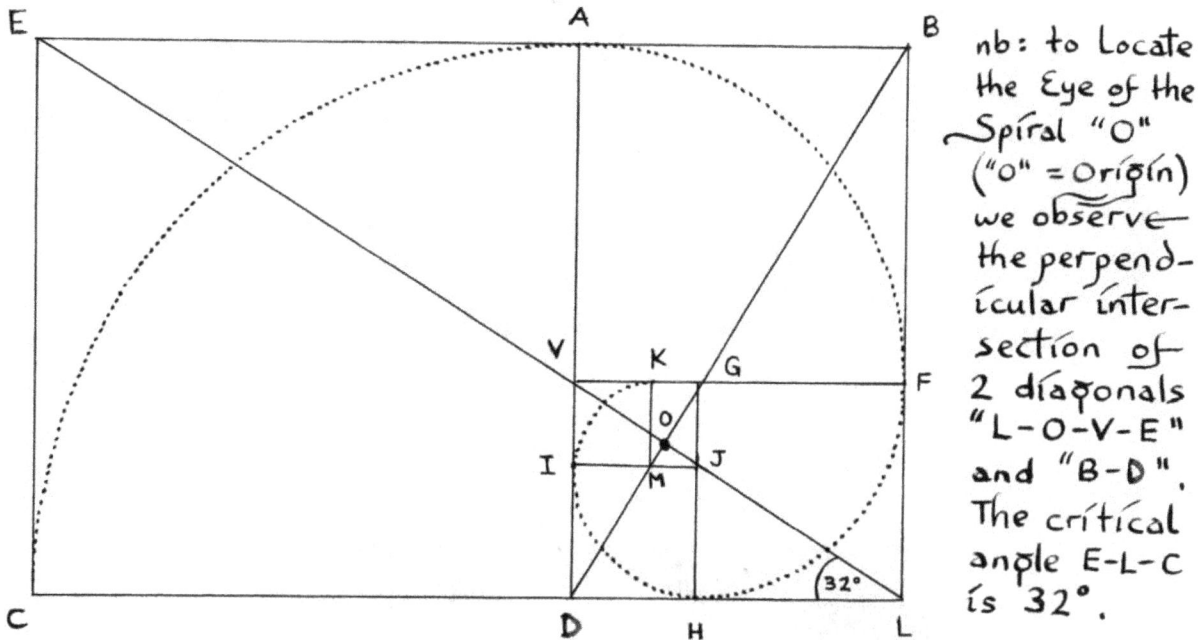

nb: to Locate the Eye of the Spiral "O" ("O" = Origin) we observe the perpendicular intersection of 2 diagonals "L-O-V-E" and "B-D". The critical angle E-L-C is 32°.

Fig 5

The Fibonacci Numbers can be represented as Squares.
Each square produces a quarter-arc circle.
The overall bounding shape that contains the Golden Mean Spiral
is the Golden Phi Rectangle.

This is also known as the Logarithmic Spiral, which is typical of shell growth. Successive stages of growth are marked by "whirling squares" and golden rectangles can be seen growing in harmonic progression from centre "O" outwardly.

To locate the eye of the Spiral "O" we observed the perpendicular intersection of 2 diagonals: L-O-V-E and B-D. The critical angle E-L-C is approximately 32°.

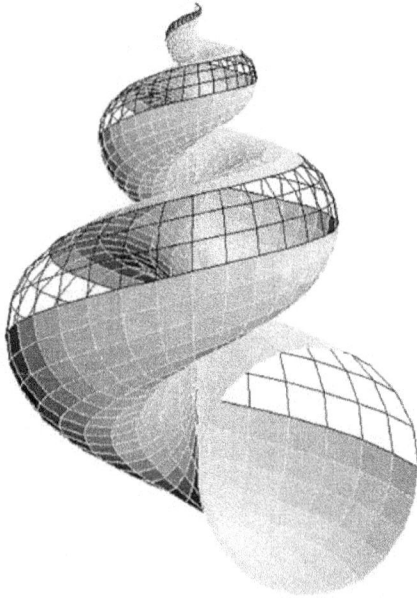

Fig 6

If you were to visualize each flat square

of the Fibonacci Numbers as a cube,

you would produce a 3-Dimensional form

of the 2-Dimensional flat spiral.

Fig 6a

When we convert these Fibonacci Numbers
into 3-Dimensional cubes, the resulting curve
fits snuggly into the familiar sea-shell patterns.

SPIRAL EXTENSION OF HELIX

using the Golden Rectangle as
geometric basis of a spiral.
Sequences the ratio: 1.618033989
Golden Mean (Φ) (phi).
in perfect length progression as Φ
in perfect area progression as Φ^2
in perfect volume progression as Φ^3
CONSERVED RATIO. CONSERVED MOMENTUM.
CONSERVED IN-FORM-ATION. CONSERVED MIND.

IN THIS IMAGE, THE DIVINE PROPORTION, GOLDEN
MEAN PROGRESSION OCCURS IN THE INCREASE
OR DECREASE IN SIZE OF UNITS- THE 1.618033989
RELATIONS IN DIMENSIONS OF VERTICAL TURN,
HORIZONTAL TURN, AND RADIUS.

VERTICAL TURNS X/Y = Y/Z = GOLDEN MEAN

HORIZONTAL TURNS AB/BC = BC/CD = GOLDEN MEAN

RADIUS OF TURNS OA/OB = OB/OC = GOLDEN MEAN

RATIO OF THE SQUARES LINES = Φ

RATIO OF THE SQUARES AREAS = Φ^2

RATIO OF THE CUBES VOLUMES = Φ^3

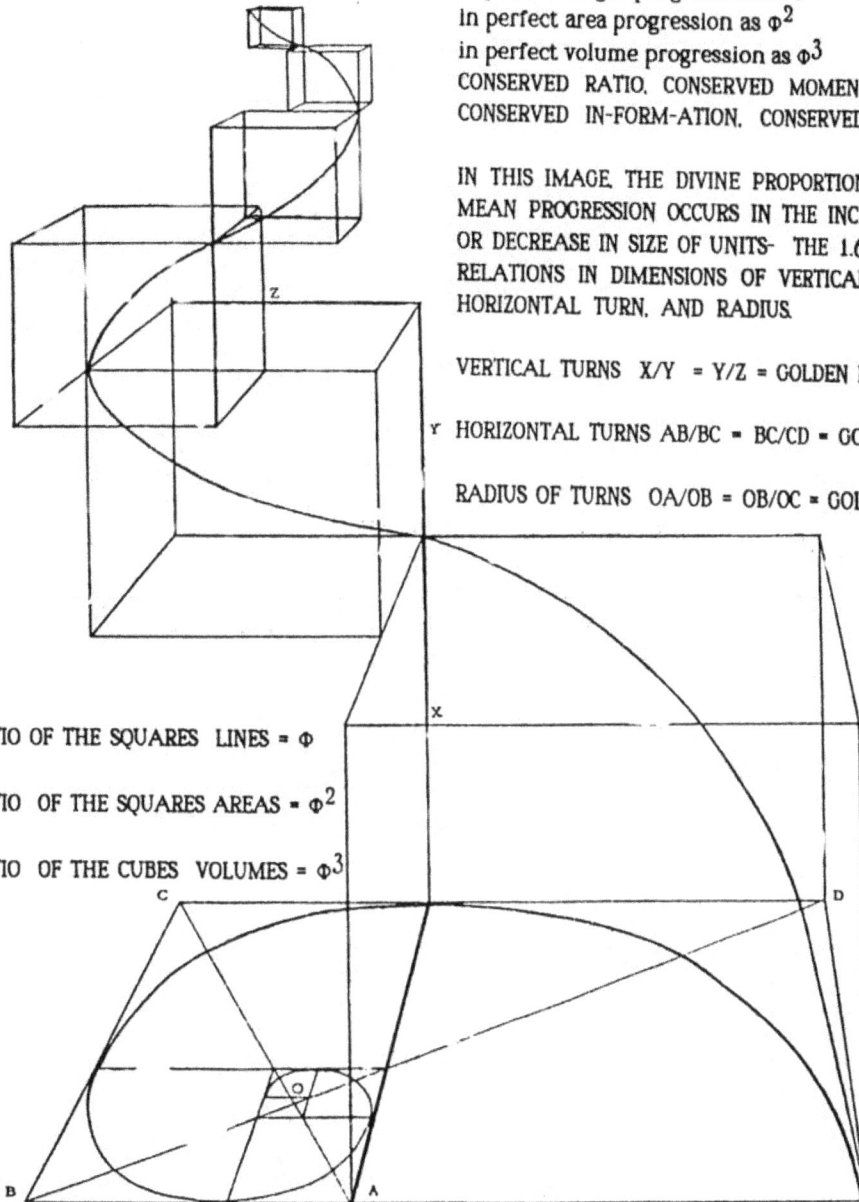

Fig 6b

The progression from the 2-Dim Nautilus Shell Spiral to the 3-Dim Ram's Horn. (as shown by Ann Tyng and reprinted in one of Dan Winter's earliest books called: "Planet Heartworks" in 1988).
Visualize each cube having an arc going from one corner to the furthest and opposing corner to generate this smooth spatially correct curve.

C~onstant~ = 1.08

(This following information was first heard on: The documentary on Numbers called "The Code", narrated by Professor Marcus du Sautoy, in the UK. (www.bbc.co.uk/code). Its 3 hours long, in 3 parts, 1 hour long each.

On the second film, at 45 mins, and 36 secs along, he is holding the nautilus shell that has been sliced through its longitudinal middle (see diagram Fig 6a above), showing all the chambers that grow each year, and he begins to measure each chamber in length with a digital ruler tool to decimal places, and presents the following data to indicate that there is a constant rate or state of growth. He divides each chamber length by the previous year's chamber length and keeps arriving at the same rate of proportional growth. The answer is repeatedly **1.08** !!!

Here is the numerical data given by Jain, starting with idealized length of 1. (On the documentary, the first figure given was 3.07, followed by 3.32, and it can be seen that 3.32 ÷ 3.07 = 1.08).

CHAMBER LENGTHS OF NAUTILUS SHELL REVEALING THE CONSTANT OF 1.08 starting with idealized length of 1		
1	1.08 ÷ 1	= 1.08
1.08	1.1664 ÷ 1.08	= 1.08
1.1664	1.259712 ÷ 1.1664	= 1.08
1.259712	1.36048896 ÷ 1.259712	= 1.08
1.36048896	1.469328077 ÷ 1.36048896	= 1.08
1.469328077	1.586874323 ÷ 1.469328077	= 1.08
1.586874323	1.713824269 ÷ 1.586874323	= 1.08
1.713824269		

Fig 7

Each successive stage of the Nautilus Shell has a rate of growth of 1.08

Unfortunately, over time, this sacred symbol of the 3-Dimensional Phi Spiral, as most pagans know, was demonized and found its place as 2 twisting horns on the head of the notorious Christian Devil to stain the Sacred Spiral's divinity with Fear. Fear means that no access to this Memory can be achieved which keeps the masses in the Dark Ages, until Consciousness begins to awaken, as is happening now.

(F.E.A.R. also is an acronym for **F**alse **E**vidence **A**ppearing **R**eal).

Regarding the Demonization of the Pentacle, having placed an inverted pentagram on the Devil's Third Eye, most people understand that the divine symbol of the Witches Pentacle (see Fig 8b) has also been smeared:

Fig 8a	Fig 8b

The Pentacle, has over 200 expressions of 1:1.618 hidden within its simple geometry. Where any two lines intersect, the proportions happen to be Fibonacci Numbers. When you draw a smaller Pentacle inside the larger Pentacle, it reduces or compresses at the rate of 1.618. And so on forever, diminishing into the micro atom or enlarging to the macro universe.

Fig 8b shows the traditional 15th Tarot card from the Fez Moroccan deck, illustrating 2 people chained to the forces of the reversed or inverted Pentagram Star seen at the crown chakra of the demonized Pan, originally the Greek God of Nature & Countryside.

The Pentacle is considered the Male Aspect of Sacred Geometry, as it consists of straight lines. Whereas, when we view yantras or power diagrams that involve curves, like Fig 2 above, it is called the Female Aspect.

The ultimate symbol of the Phi ratio is The Infamous Pentacle.

All living proteins are pent shaped.

The Dodecahedron is the 3-Dimensional form of the 2-Dimensional Pentacle, the true shape that all Pagans must resonate to is known as the 5th Element of Ether or Spirit, the most important of the 5 Platonic Solids known as the **Dodecahedron** ("do" means "2" in Greek, and "dec" means "10", as in decimal or decade, and "hedron" means face. Thus it is a 12-faced polyhedron with 12 pentagonal sides where all lines and all angles and all faces are equal and all 20 vertices touch the sphere).

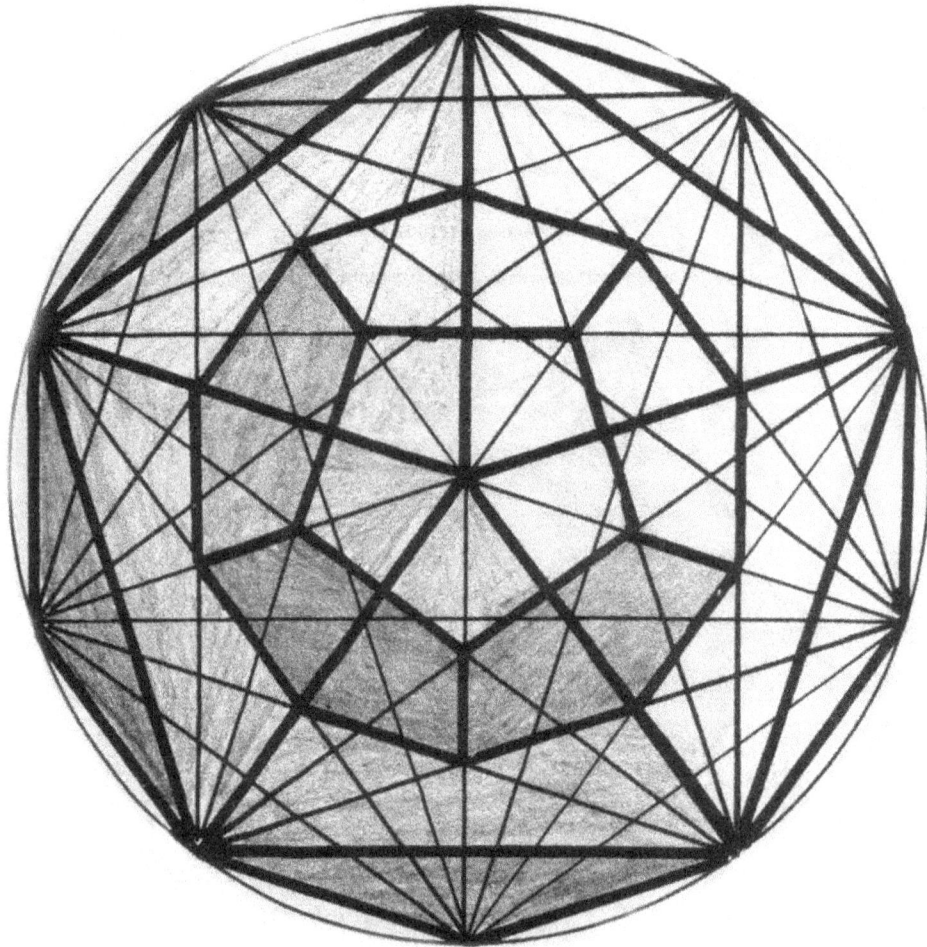

Fig 9

Infinite 1.618 Series of the Dodecahedron (12 Pentacles) and Icosahedron (20 Equiangular Triangles) nesting relationship.

The diagram Fig 9 here shows the Infinite Series of the Dodecahedron, when stellated forms a larger Icosahedron of 20 triangular sides, and when this is stellated, the icosahedron forms the dodecahedron. Thus the infinite process continues, from the atom to the galaxy, the size or shape changes, which is the key to Alchemy, but what remains the same is the wavelength of 1.618033.

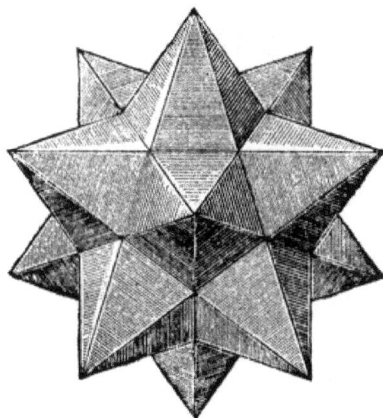

Fig 9a

The Stellated Dodecahedron, showing the 12 pentagonal pyramids. The stellations must be in the ratio of 1: 1.618... for the magic to happen, so that the joining of the 12 apices or tips or vertices forms the Icosahedron.

This is called **Scale Invariancy**, which means that size does not matter, what is is important is the scale or unchanging proportion.
Also, the 10 sided diagram in Fig 9 means that the 3-Dim Dodecahedron has a 2-Dim decagonal shadow or view, in the jargon of the sacred geometrician, it is shortened to "the dec view of the Dodeca".

Most people know for example that the **Mastercard** fits snuggly into the square of your hand, when you hold it, but if you examined the mathematics of this shape, it is a Phi Rectangle exactly 21:34.

The list of where Phi occurs in Nature, in Crystal, in Space, is endless.
After a while of studying these proportions, you begin to see it everywhere, as you walk through your town or forest. Ensure that your children learn the essence of this Living Mathematics of Nature.

PART 2

WHAT YOU DON'T KNOW
ABOUT PHI and THE FIBONACCI SEQUENCE:

JAIN'S DISCOVERY of
The REPEATING 24 PATTERN

There is much knowledge about the Omniscient Phi Ratio, but what you don't know is also of great interest:

We have been told for thousands of years that this Phi Ratio is an infinite decimal that has no end and no visible pattern, (which is correct), that it's vibration of **1.1618033988** just keeps going on like a non-sense number, like the famous Pi of 3.141592 on and on forever without any symmetry (which is incorrect, as there does exist an invisible yet secret pattern). Conventional mathematics takes you to a certain ceiling or level of understanding, but the mathematics that is about to be revealed, is a higher level of Knowledge that will lift this ceiling of forgetfulness.

Our eyes need to be sharpened to develop a sense of X-ray eyes that are capable of distinguishing the inherent Order amidst the dominating Chaos.

But the truth is, we have been duped, or dumbed down by the conventional learning institutions. The very definition of the Golden Mean or Divine Phi Ratio is that there is a pure and marvelous symmetry embedded in the numbers, shared wavelengths that can travel from the atom to the universe without being self-destructed, in a sense that are immortal. Only the 1.618 ratio is fractal enough to ensure survival, as it knows how to be **self-similar, embeddable**, it knows how to be **recursive** in the micro and macro. That is, there must be a pattern visibly hidden somewhere in this infinite number. A useful image here is that of the nested Russian dolls, one within the other, from the small to the large, as a key to the DNA molecule and the spiral galaxies.

So let us say that everyone knows of the Fibonacci Sequence, but what you don't know is how this sequence relates to Time Travel and is a true Time Code.

There are 16 Vedic Mathematical Sutras, and one of them is known as "Digital Compression" which simply means we "add the digits", thus when we look at the Fibonacci Numbers: 1, 1, 2, 3, 5, 8, we notice that these are all single digits, which is our aim to express all the following digits as singularly reduced digits. Thus when we come to the next number which is 13, we add the digits: 13 = 1 + 3 = 4, and when we come to the next number which is 21, we add the digits: 21 = 2 + 1 = 3 etc. When we have effectively reduced the Fibonacci Sequence into single digits from 1 to 9, we see that it now appears like this:

1, 1, 2, 3, 5, 8, 4, 3, 7, 1, 8, 9, 8, 8, 7, 6, 4, 1, 5, 6, 2, 8, 1, 9

Fig 10
The 24 infinitely recurring singly-reduced digits of the Fibonacci Sequence

This is still a meaningless sequence of 24 numbers but it begins to make sense when we examine and reduce the next 24 Fibonacci Numbers and the next set of 24. It happens quite magically that the numeric reduction of the Fibonacci series produces an infinite series of 24 repeating digits:

If you take the first 12 digits and add them to the second twelve digits and apply numeric reduction to the result, you find that they all have a value of 9.

The PHI CODE of 12 COMPLEMENTARY PAIRS OF 9													
1st Set of 12 Numbers	1	1	2	3	5	8	4	3	7	1	8	9
2nd Set of 12 Numbers	8	8	7	6	4	1	5	6	2	8	1	9

Fig 11

The 12 Complementary Pairs of 9 in the Phi Code,
Expressing the galactic mathematics of Base 12 and Base 9.

These 12 pairs of 9 have a sum of **108**. (Is it mere coincidence that the external angles of the Pentacle are 108 degrees!). It means that this frequency of 108 is the hidden pulse, or rhythm that is the essence of the living mathematics of Nature, it is the reason why the Vedas worshipped this number by incorporating it into their 108 rosary beads when they chant their most famous of all mantras, The Gayatri mantra, the most famous Eastern song or Prayer of Enlightenment which is always chanted 108 times, also has a rhythm of 24 syllables! What else embodies mathematically these two numbers 24 and 108. It can only be the Reduced Fibonacci Sequence.

Jain's Lost Secrets of The Phi Code includes a lesson in x-raying the Fibonacci Sequence: 0, 1, 1, 2, 3, 5, 8, 13, 21, 34, 55, 89, 144 etc to discover where the infinite phi expression has its underlying rhythm, a distinct periodicity of 24 recurring digits that can not be seen by the Western Mathematical Eye but only by the Truth of Vortex Mathematics expressed as Vedic Mathematics!

Really this Knowledge predates the Vedas and Atlantis and is better referred to as Galactic Mathematics.

This material shakes the whole foundations of the current Western mathematics curriculum, will rewrite our current world view on Mathematics and will require a new definition of the Golden Mean: that its decimal does carry on infinitely but within its infinity there is recursive or self-similar beauty. Scientists like Naudin of France and Metaphysicians like Marco Rodin wind toroidal coils in sync with this recursion or rhythm or pulse or pattern of 24 digits to get more output than their input!

Just remember that ultimately we need to ask ourselves, what shape encodes this quality of 24, what shape has 24 faces and 24 edges. There is only one shape that does this, and it is the atomic structure of Silicon Chip which is a Star

Tetrahedron, or two interpenetrating or inter-digitating Tetrahedrons, that forms the basis of many crystals. When you join the 8 vertices of this shape, you get the Cube.

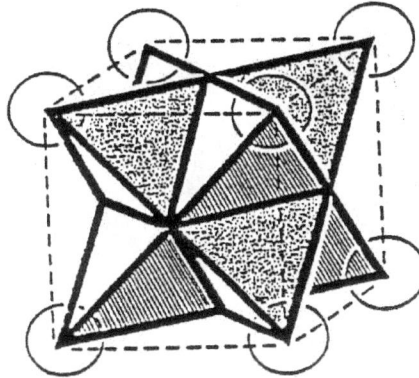

Fig 12

The Star Tetrahedron, (having 24 Faces and 24 Edges)
which is the 3-Dimensional form of the Star of David,
is the geometric equivalent to the Phi Code expressing precisely the
24ness exhibited in the Reduction or Compression
of the Fibonacci Numbers.

Basically, **Shape stores Memory.** Due to this profound mathematical discovery, there is now established a mathematical and spiritual connection between Technology and Consciousness, between Sacred Geometry and Inter-Dimensional Time Travel. This Number 24, this StarGate, this clock of 24 Hours, is what Einstein called the Fourth Dimension: Time.

If this number "24" cropped up again in another Mathematical Sequence, would you be convinced that this is no mere co-incidence.
According to a German chemist, Dr Peter Plichta, the Egyptians knew the secret to the hidden pattern of Prime Numbers. Imagine if you wrote the numbers from 1 to 24 in a circle, going clockwise, and the next 24 numbers concentrically around it, and you repeated these many rings of 24 consecutive numbers, have a guess where all the prime numbers would lie? You guessed it, on 4 distinct diagonals or diameters that form the 4th Dimensional Templar Cross.

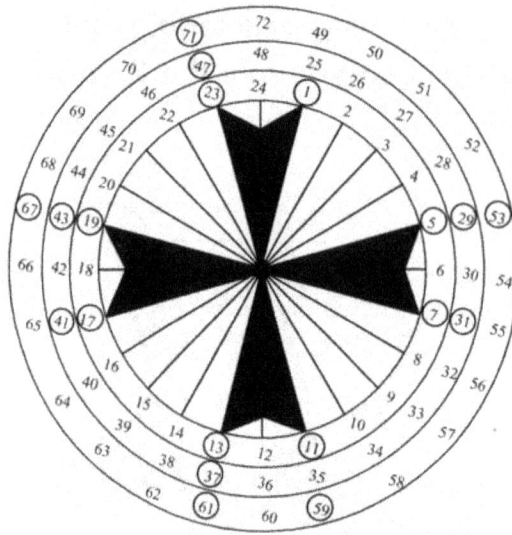

Fig 13
**The Symmetry of the Prime Numbers
shows the 4th Dimensional Templar Cross
based upon concentric rings of 24 consecutive numbers.**

Yet again, all the academic books are embarrassingly wrong, they educate our children that the Prime Number Series has no meaning nor pattern. I wonder why our military forces use the higher end spectrum of these "Twin Primes" (like 17 & 19) of the Prime Number Series for advanced encryption technology.

For those readers who are aware of the Philadelphia Experiments that aimed to turn military naval boats invisible during times of war, did you know that the science behind it was purely working with the Fibonacci Numbers and magnetic fields.

Did you know that the total success of the modern computer age is dedicated to the Fibonacci Numbers, for the ability to send large files, from one system to another, anywhere in the world, depends on the ability to compress files? (which is what the sunflower does with its seeds, they compress into the 21:34 ratio, otherwise, if the seeds along the counter-rotating spiral arms were arranged say equally at 21:21, they would simply fall out of their floret). Thus computer designers, by emulating Nature's Pure Principle, solved their problems thanks to what we call **Fractal Compression**, which as a frequency of numbers is the pure **1.618033988...** consciousness. Remember that Phi is not a Number, but rather a "cascading of frequencies" of the Fibonacci Numbers.

Yet why are we not educating our children about this fountainhead of Knowledge?

I have been commissioned all over Australia and globally to teach children, over the last 20 years, with one simple job: to teach the Beauty of Mathematics.

Its quite an evolution of new mathematical gems coming through, and if you stay in tune with this website, www.jainmathemagics.com I will have more of these essential and psycho-active diagrams posted for you. These articles will help you discover that Maths is Art, Maths is Science and Maths is History. They will help you to explore Sacred Geometries that are invisibly constructed and nested within the Heart, and ultimately teaching you that all Knowledge is within You. You only need to learn how to **Remember** this Knowledge, it is already there.

The most important element in Sacred Geometry is that we already know this information, that the process is about remembering this Lost Knowledge of the PHI CODE that is already morphologically and geometrically encoded in your bio-magnetic and bio-electrical fields.
Essentially, the human body is in resonance with the **Living Mathematics of Nature.** Knowing this, is part of your Ascension Process.

Ultimately, this Pythagorean Knowledge is not an intellectual process, but rather that these essential geometries are invisibly constructed and nested within the Heart as the Centre of a greater Galactic Grid connecting us to all Memory of all Universes and Atomic Levels. Learn to sit still everyday, and in your daily meditations there is one true key to mastery, and that is to give thanks for all that we already have: give Gratitude.

Fig 14
**Transverse cut of the Torus, showing the nesting of many Tori,
and separated by distances of the Fibonacci Numbers.
The Wormhole Centre that bridges the Worlds is your Heart.
This material is classified as the Advanced Knowledge on the
12:24:60 Phi Code, but all Knowledge is for Sharing,
when the Time is right.**

In the next article/lecture, we will learn the advanced and practical use of integrating this ancient Knowledge of the Infinitely Repeating 24 Phi Code by breathing in the Memory of the Celestial (that which is above), and the Terrestrial (that which is below) into our Heart. The meeting place of these counter-rotating fields, like the dynamics of the Sunflower floret witnessed in Nature, is One Breath meeting in the Heart. A One Breath Heart Meditation, copying and imitating Nature, is the fulfillment of this intellectual knowledge so that it becomes an Experience, an unlimited Journey through Time and Space, navigating with the powerful tools of Numbers, Fixed Eternal Design. This Meditation, the key subject that I teach around the world, and now to Children, is called the **EARTH-HEART Meditation**.

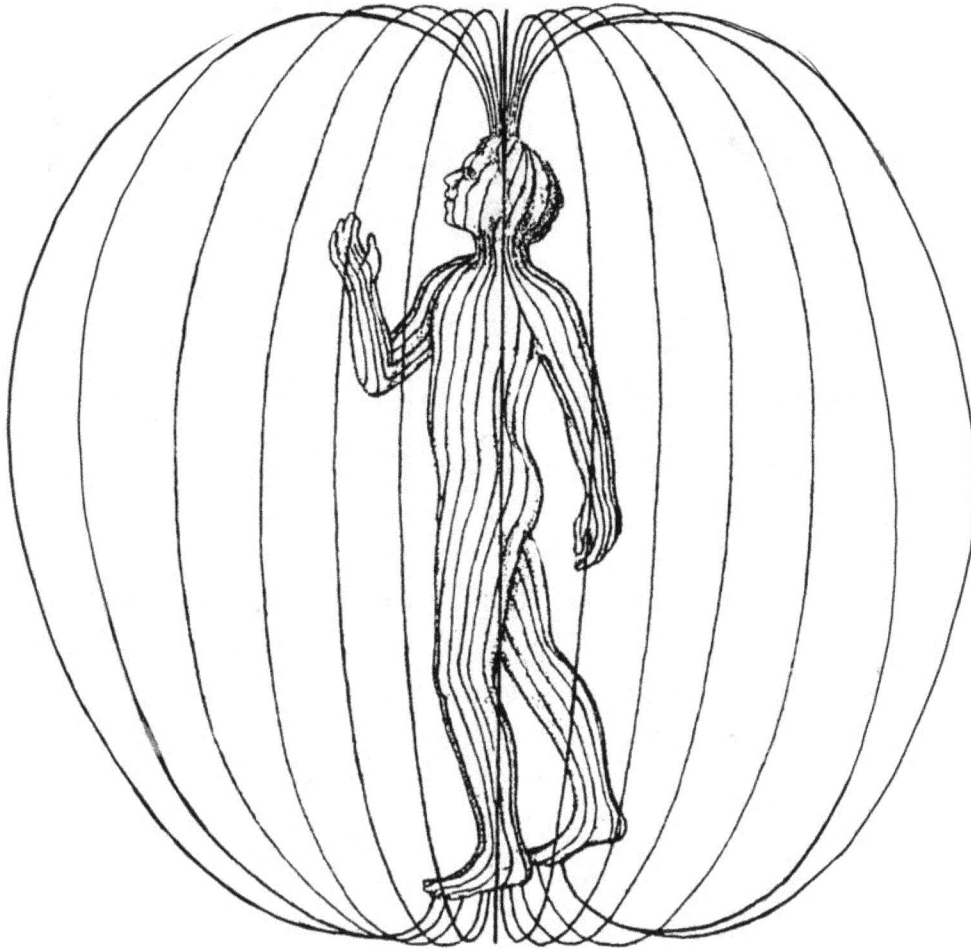

Fig 15
**Jain's Earth-Heart Meditations
is the embodiment of this Ancient Knowledge,
by simple, effortless in-breaths into the Heart,
and out-breaths to your surrounding domain, expanding to the
spherical diameters of the Fibonacci Sequence.**

PART 3

WHAT YOU NEED TO KNOW ABOUT PHI and The FIBONACCI SEQUENCE:

TIME CODE 12:24:60 ENCRYPTED in the FIBONACCI SEQUENCE and PASCAL'S TRIANGLE

Having established that all the books are wrong and there is indeed immense symmetry in the Fibonacci Sequence, specifically an infinite repeating pattern of 24 reduced digits, it is in this section that we stumble upon another important cycle of repeating digits. Before I give it to you, you must ask yourself:

"What number or numbers, symbolically, represent 'TIME' to you?"

Many people will say, 12 or 24 hours or 60 seconds to the minute or 60 minutes to the hour, which is the vestigial remains of the Babylonian use of Base 60 several thousands of years ago etc

What if, in this article, I could link these critical numbers:

12 : 24 : 60 in a manner that connects them to this concept of Time Codes to Nature. Would you be impressed?

In Part 2, I showed how the number 24, symbol of Hours in the Day, is hidden in the Fibonacci Sequence, thus this section of Part 3 will now focus on where the Cycles of 60 reside.

"The Cycle of 60 Pattern" in the Fibonacci Sequence:
I am rewriting the same sequence again, but this time I will highlight in bold only the **final digits** of the Fibonacci Sequence:

0, **1**, 1, **2**, 3, **5**, **8**, 13, 21, 34, 5**5**, 8**9**, 14**4**, 23**3**, 37**7**, 61**0**, 98**7**, 159**7** ...

Fig 16
The Final Digits of the Fibonacci Sequence

1	1	31	1346269
2	1	32	2178309
3	2	33	3524578
4	3	34	5702887
5	5	35	9227465
6	8	36	14930352
7	13	37	24157817
8	21	38	39088169
9	34	39	63245986
10	55	40	102334155
11	89	41	165580141
12	144	42	267914296
13	233	43	433494437
14	377	44	701408733
15	610	45	1134903170
16	987	46	1836311903
17	1597	47	2971215073
18	2584	48	4807526976
19	4181	49	7778742049
20	6765	50	12586269025
21	10946	51	20365011074
22	17711	52	32951280099
23	28657	53	53316291173
24	46368	54	86267571272
25	75025	55	139583862445
26	121393	56	225851433717
27	196418	57	365435296162
28	317811	58	591286729879
29	514229	59	956722026041
30	832040	60	1548008755920

Fig 16a
The first 60 Final Digits of the Fibonacci Sequence

This time, I will only write out the final digits:

0, 1, 1, 2, 3, 5, 8, 3, 1, 4, 5, 9, 4, 3, 7, 0, 7, 7 ...

Fig 17

**Beginning Sequence for the final digits
of the Fibonacci Sequence
displaying a distinct Periodicity or Cycle Length of 60.**

There is no point asking you to observe if you can see a pattern as it turns out that the series is **60** single digits long, and that these **60 digits keep repeating forever** and forever in that same series. In mathematics, we say that the series of final digits has a **periodicity of 60** or that it has a distinct cycle length of 60.

It means, that this recursive cycle repeats itself after Fib(60) which is: 1,548,008,755,920 (approximately 1.5 trillion).

So here it is, another hidden, infinite yet simple code based on 60.

Just for the record, here is the string of 60 final fib digits, but arranged in a 5 x 12 table.

There may be other hidden patterns lurking within this table, or other arrangements of the same data, eg: you could have these 60 final fib digits, arranged in a 6x10 array or a 4x15 array or a 2x30 array.

nb: Zeroes are used in this table, as one of the Fibonacci numbers F(15) is "610" and this final digit of "0" must be recorded as part of the series.

1	1	2	3	5	8	3	1	4	5	9	4
3	7	0	7	7	4	1	5	6	1	7	8
5	3	8	1	9	0	9	9	8	7	5	2
7	9	6	5	1	6	7	3	0	3	3	6
9	5	4	9	3	2	5	7	2	9	1	0

Fig 17a

The 60 final digits of the Fibonacci Sequence, in a 5x12 rectangular array,

What pattern do you notice?
Can you see that the 3rd, 6th, 9th and 12th column contains all the Even Numbers, (the sum of each column is 20),
and all the other columns are comprised of Odd Numbers (the sum of each column is 25).
This gives a total sum, for all 60 digits, to be (4x20 + 8x25) = 280

Lets have a look at the 4x15 array:

The Last 60 Final Digits of the Fib Seq: ARRAY 4 x 15														
1	1	2	3	5	8	3	1	4	5	9	4	3	7	0
7	7	4	1	5	6	1	7	8	5	3	8	1	9	0
9	9	8	7	5	2	7	9	6	5	1	6	7	3	0
3	3	6	9	5	4	9	3	2	5	7	2	9	1	0

Fig 17b

The 60 final digits of the Fibonacci Sequence, in a 4x15 rectangular array.

Notice how the last vertical column displays all zeroes, and the 5th and 10th columns display all 5s. All the other columns have sums of 20 and like previously, each column is composed of either Odd or Even Numbers.
(The uninteresting sums of the 4 rows in order from top to bottom are: 56, 72, 84, 68).

1	1	2	3	5	8	3	1	4	5
9	4	3	7	0	7	7	4	1	5
6	1	7	8	5	3	8	1	9	0
9	9	8	7	5	2	7	9	6	5
1	6	7	3	0	3	3	6	9	5
4	9	3	2	5	7	2	9	1	0

Fig 17c

The 60 final digits of the Fibonacci Sequence, in a 6x10 rectangular array.

Notice that 2 columns have a sum of 20 and all the other columns have a sum of 30. There is partial symmetry in the distribution of specific numbers in each column.

Let us now examine this 60 Cycle in the 3 x 20 Rectangular Array below:

The Last 60 Final Digits of the Fib Seq: ARRAY 3 x 20																			
1	1	2	3	5	8	3	1	4	5	9	4	3	7	0	7	7	4	1	5
6	1	7	8	5	3	8	1	9	0	9	9	8	7	5	2	7	9	6	5
1	6	7	3	0	3	3	6	9	5	4	9	3	2	5	7	2	9	1	0

Fig 17d

The 60 final digits of the Fibonacci Sequence, in a 3x20 rectangular array.

(For the record, there appears to be no real symmetry besides the repetition of each column being seen twice exactly or as an anagram ie: rearrangement).

Let us examine this 60 Cycle in the 2 x 30 Rectangular Array:

The Last 60 Final Digits of the Fib Seq: ARRAY 2 x 30																													
1	1	2	3	5	8	3	1	4	5	9	4	3	7	0	7	7	4	1	5	6	1	7	8	5	3	8	1	9	0
9	9	8	7	5	2	7	9	6	5	1	6	7	3	0	3	3	6	9	5	4	9	3	2	5	7	2	9	1	0

Fig 17e
The 60 final digits of the Fibonacci Sequence, in a 2x30 rectangular array.

Notice that 28 of the 30 columns have a sum of 10 and all the other 2 columns have a sum of 0.

So next time someone asks you: "What is the Time?" you can present to them the "60"pattern hidden within this Fibonacci Series.
But it gets even better.

What if we were to examine all the **final two digits** in the Fibonacci Sequence and inspect for another pattern. If there is one, what would its periodicity be? That means I need to rewrite the sequence and highlight the final two digits:

0, 1, 1, 2, 3, 5, 8, 13, 21, 34, 55, 89, 144, 233, 377, 610, 987, ...

Fig 18
The final two digits in the Fibonacci Sequence has a Periodicity of 300

Remarkably, there is yet another distinct recursion happening, but since it is too long to write out, I will just state that it happens every 300 digits long. That means after Fib(300), the final two digits keep repeating the same sequence again, then again then again, forever. The periodicity or cycle length is 300. (Before I give you more hidden patterns, at this stage, I would like you to note the relationship between the two cycles of 60 and 300. We can summarize that as a proportion, they are in a 1:5 ratio).

Let us continue. The fact that we have discovered two incredible patterns in the final digits and then with the two final digits, you could easily surmise now that there may be a good chance that there exists patterns in:
the **last three digits** and the **last four digits**, and the **last five digits**.
Again, the Patterns are clearly there, thanks to the help of modern computers:
It can be realized that there are more and more patterns:

a) For the last **three** digits, the Periodicity or Cycle Length is 1,500

b) For the last **four** digits, the Periodicity or Cycle Length is 15,000 and

c) For the last **five** digits the Periodicity or Cycle Length is 150,000

and so on...
Let us make a chart of this newly discovered information, analyzing the number of Final Fibonacci Digits and the Periodicity or Cycle Length of each:

Number of Final Fibonacci Digits	Periodicity or Cycle Length
1	60
2	300
3	1,500
4	15,000
5	150,000

Fig 19
Chart of the Final Digits in the Fibonacci Sequence and their corresponding Periodicities or Recursion Cycle Lengths

Are there other obvious patterns?

What is interesting is the Ratio or Proportion of each of these Periodicities, and how they will relate to Pascal's Triangle, but before the Proportions are revealed, let us look at another hidden coding:

The following diagram (Fig 20) shows the traditional Pascal's Triangle and then we will show The Fibonacci Sequence disguised within its array and also the proportional numbers that appeared in Fig 19.

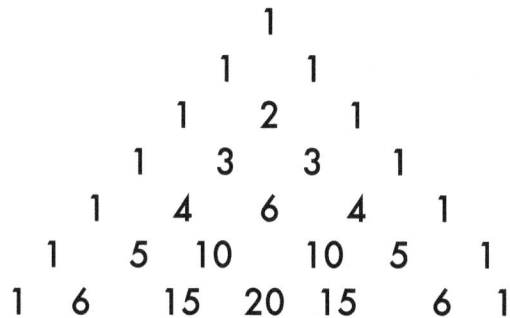

```
                1
             1     1
          1     2     1
       1     3     3     1
    1     4     6     4     1
 1     5    10    10     5     1
1   6     15   20    15    6    1
```

Fig 20
Pascal's Triangle

Since we are avid Pattern Hunters, we want more patterns and links, as now it is becoming quite a serious matter, that Nature is not random, but obviously there is some inherent Order in how the Universe operates. One thing that cannot lie, is Proportion, as it is a forever science or a universal language.

Let us therefore list these important numbers, and examine their proportions to one another:

60 : 300 : 1,500 : 15,000 : 150,000

Let us now compare each number to the one adjacent to it:

60:300 = 1:5

300:1,500 = 1:5

1,500: 15,000 = 1:10

15,000:150,000 = 1:10

To continue this investigation, in our attempt to link these ratios to other important mathematical sequences, we will list the above ratios in this format:

1 : 5 : 10 : 10

And what if the fundamental basis for this underlying or hidden mathematical structure was known by the ancient Vedic scholars.

If the Indian scholars from 2,000 years ago knew of this Fibonacci Sequence, would we by right acknowledgement be forced to change the name to the original discoverers of the Fibonacci Code. According to Vedic texts, the Fibonacci Sequence was attributed to HemaChandra-Gopal well before it arrived to Europe via Arab wars and traders. But in all my research, I have not seen any real or clear evidence about Fibonacci Numbers in the Vedas.

Fibonacci was living in the time of the 12th Century, Europe, and HemaChandra-Gopal lived about 2,000 years ago in Bharat or India. Do we now call the Fibonacci Sequence the HemaChandra-Gopal Code. Or do we just accept that there were other pre-Vedic cultures, like Babylonia, Sumeria, etc (and master alien races from other star systems) and that all Knowledge has existed, and no-one can ever really put their name to any discovery which is only a Re-Discovery. Thinking along these lines, there is another famous pattern called Pascal's Triangle, but there are claims that it also has its so-called origins in ancient India.

The famous **Meru Prastera** Pattern is also mistakenly called Pascal's Triangle. (The image of "Meru" is like a holy mountain, like Mt Olympus is to the Greek Gods).

```
                        1
                1               1
          1           2               1
       1          3           3           1
    1        4          6          4         1
  1       5         10         10         5         1
 1       6         15         20        15        6        1
```

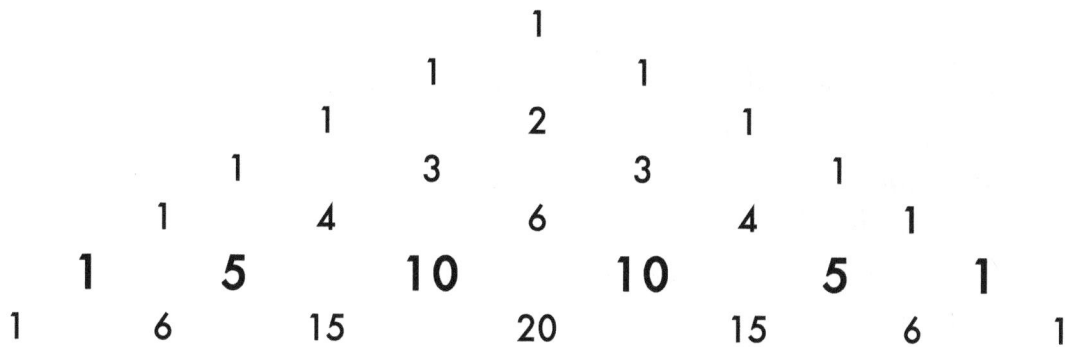

Before explaining what Pascal's Triangle is all about, and how to create it, let me immediately connect the previous code of 1 : 5 : 10 : 10 to be found in the above Pascal's Triangle, along the 6th horizontal line (as shown above in Fig 20a).

Such revelations basically connects the Living Mathematics of Nature, which is the Fibonacci Sequence, to other mathematical giants of Inherent Order and Underlying Symmetry like Pascal's Triangle above.

All is connected.
All is a Unified Field.
All is One.
These two important sequences establishes a direct connection to the **12 : 24 : 60** Galactic Time Code, the **24 Gods that are One.**

TIME CODES 12:24:60
Or
THE TWENTY-FOUR GODS THAT ARE ONE

I ask You, the audience:

What numbers, symbolically, represent "Time" to you?

Please call out any numbers that you know to be relevant.

Many people will say, 12 or 24 hours or 60 secs to the minute etc

What if, in this lecture that I could link up to you, these critical numbers:

12 : 24 : 60

in a manner that connects them to this concept of The underlying Time Code to Nature, (and with the Physics of Time Bending) with Nature being symbolized by the Divine Proportion 1:1.618...

Would you be impressed?

And what if the fundamental basis for this underlying or hidden mathematical structure was known by the ancients Veda scholars and represented as the HemaChandra-Gopal Code that we now mistakenly call the Fibonacci Sequence! But it was most likely known by the lost civilizations of ancient Atlantis and Lemuria which predated the Vedic race, so it is universal knowledge, it belongs in the Public Domain, not to any one race. For this book only, for simplicity sake, let us call it "Pascal's Triangle".

<u>By definition, how to create Pascal's Triangle</u>:

Each entry in the triangle is the sum of the two numbers in the row above. A blank space can be taken as "0" so that each row starts and ends with "1".

Thus you can see that "1 + 1" = "2" and

"1 + 3" = 4 etc

In fact, most PhiBonatics already know that the diagonals in Pascal's Triangle have a sum relating to the Fibonacci Numbers.

The Fibonacci Numbers in Pascal's Triangle

```
1
1   1
1   2   1
1   3   3   1
1   4   6   4   1
1   5   10  10  5   1
1   6   15  20  15  6   1
```

Fig 21

Re-arrangement of the Meru Prastera Triangle
All rows are now Left-Aligned.

**Notice how the Diagonals have sums
which are the Fibonacci numbers?**

This time I am going to highlight two parallel diagonals in bold so that you can begin to see the first pattern:

```
1
1   1
1   2   1
1   3   3   1
1   4   6   4   1
1   5   10  10  5   1
1   6   15  20      15  6   1
```

Fig 21a

**The Diagonals of the Pascal's Triangle
create the Fibonacci Numbers.**

Highlighting the Diagonals to reveal the Fibonacci Sums
Can you see in bold that 1 + 3 + 1 = **5**
Can you see in that 1 + 4 + 3 = **8**
Can you see in bold that 1 + 5 + 6 + 1 = **13**

The numbers of the diagonals continue like this forever! Always adding to a Fibonacci Number, in order.

In mathematics, it is important to submit these findings as a Mathematical Paper, so others can share it. Thus, the general principle that we have just illustrated here is:

"The sum of the numbers on one diagonal is the sum of the numbers on the previous two diagonals."

Another arrangement of Pascal's Triangle. Perhaps a clearer way of demonstrating the same principle is to have it represented more linearly, having the numbers simply tabulated one under the other, but with one change, by sliding all the rows over by 1 place, as shown in the table of Fig 22:

1	2	3	4	5	6	7	8	9	10
1									
	1	1							
		1	2	1					
			1	3	3	1			
				1	4	6	4	1	
					1	5	10	10	5
						1	6	15	20
							1	7	21
								1	8
									1
1	1	2	3	5	8	13	21	34	55

Fig 22 Alternative diagram illustrating how the Rows or Diagonals of Pascal's Triangle create the Fibonacci Numbers. The Bottom Row in bold is the sum of the digits posited in the vertical columns.

OTHER INTERESTING FACTS
ABOUT PASCAL'S TRIANGLE

Meru Prastera Triangle has lots of uses including:

1 - To solve problems in Probability.

Imagine you were asked how many permutations of Heads (H) and Tails (T) or different ways are there of throwing 3 coins onto a surface, we know that the answer is 2 to the power of 3 which is $2^3 = 8$.

But if you were specifically asked what are your chances of getting a Head (H) in any throw how would you record your entries to fully analyze the data?

This would be the solution:

3 heads: H=3 is found in **1** way (HHH)
2 heads: H=2 can be got in **3** ways (HHT, HTH and THH)
1 head: H=1 is also found in **3** possible ways (HTT, THT, TTH)
0 heads: H=0 (i.e.: all Tails) is also possible in just **1** way: TTT

Do you notice the highlighted numbers in bold: **1 3 3 1**.

It is the 3rd horizontal row of the Meru Prastera Triangle.

2 – The Meru Prastera Triangle can also help find terms in a Binomial expansion: $(a+b)^n$
eg: $(a+b)^3 = 1a^3 + 3a^2b + 3ab^2 + 1b^3$

I will not go into the theory here. I have merely set the ground work to introduce you to the amazing worlds of the HemaChandra-Gopal / Fibonacci Sequence and the Meru Prastera / Pascal's Triangle.

CONCLUSION

What now arises for me as an adamant researcher, is the contradiction going on between this important and timeless, universal discovery, of 12:24:60 and how it conflicts with the much flaunted and bantered new time code called the Mayan Calendar whose main proposition is 13:20.

Although the current literature on the Mayan Calendar is exceptionally exciting, it must be reviewed and reconsidered. It is informally written by the late Jose Arguelles, an author and artist and visionary. Their key catch phrase is: 12:60 = TIME IS MONEY (the old paradigm that we are collectively shifting out of) and 13:20 = TIME IS ART (the new paradigm that we are realigning ourselves to).

But his premise is challenged by Dr Bernard Jenkins, illuminating all his readers that Jose's time count is wrong, and that there are not 13 full-moons in the year. Current Mayan Elders are demanding that Jose Arguelles recant and not published his imagined Day Out Of Time that he introduced into the westernized Mayan Calendar.

Actually, we need to know exactly how many full moons there are in the year. I have researched that there are not 12 nor 13, but somewhere in the middle, like 12.4. The purpose of this article, is to not believe all the literature you read. We are Pattern Hunters, looking and investigating for clues that are based on timeless laws of the Universe's cycles. Yes we know that the Fibonacci Sequence is fully laden with the cycles or periodicities of 12:24:60 and that is definitely connected to the Meru Prastera Triangle, and thus is a clue to Time Travel, Space Travel, Inter-Dimensional Travel... that is the next step, linking this Divine Mathematics to the Divine Sciences.

So I live in an area in Byron Bay which has the potential to be a new model rural and coastal town that is aiming to be self-sustainable for the future in terms of recycling our own waste, buildings with no chemicals, cleansing the water-ways, and introducing this Vedic and JainMathemagics into our schools. But all my peers are waving the Mayan Calendar flag subtly proclaiming that Base 12 is bad news, and hey man, we got to shift to the 13 frequency etc but how can I accept this bad press **when I am armed with this irrefutable evidence that Base 12:24:60 is It, was It, and will always be It.**

One of the important warnings of our life-time, I believe, is to not get caught up in the 2012 prophesies, it will become a madness in the next few years, and create so much Fear that it will globally dis-empower our Forces to bring real and effective change into this world. Since 2012 is future-based, it will only serve to keep us out of The Now, which is our real power, The Now. And Fear will be generated in every nation, that ends times are coming. Just like what happened at the turn of the bi-millennium, panic everywhere, people selling houses, Y2K bug created false fear etc. So be warned, stay in your Heart and if anyone raises the 2012 flag or banner just walk away, do not even engage in a conversation with them as they are over-inflamed and superficially correct. The wise person stays in The Now, and returns to the Garden where the true Mathematics is stored.

No doubt, the prophesies will come true, that is the changes, they will happen, but the discussion here is that Man-Woman does not know how to count Time, it has been infected with a virus; wars and many cultures have interfered with the original blueprints, so 2012 will happen, but when you are looking the other way. And all it will ever mean is a raising of consciousness at the expense of racial cleansing. Over-population must be reduced, climates must rebalance, the waters must be renewed from chemicals etc so the times they are a changing, but no calendar can predict it. If anyone talks about 2012, just walk away.

There can be no argument about it. If there were precisely 13 full moons in our year, then I would be in a position to reconsider my findings, but I can not budge.

We must keep doing the research. Go to any astronomer and enquire, how many full moons are there in the solar year, sidereal year, and how do you measure this etc... Ask other researchers and compare their findings.

The other fascinating evidence is from the book called: "CIVILIZATION ONE" by Christopher Knight and Alan Butler (www.civilizationone.com) which states quite clearly that our forbears did not compute their calendars and weights and measures based on the solar year of 365 days but rather chose to index their data against the stars, and Venus' orbit, and concluded that we actually have always used 366 days per Sidereal Year (sidereal meaning the stars). Now this is a revolution in Mathematics and Astronomy. It challenges everything you ever believed about calendrical time count. Any person or book that can shake your current belief systems is to be welcomed. We must remain open to these occasional waves of new knowledge that surface above the other waves in this ocean of infinite belief systems. But one system, above all others is a universal wave, an omniscient wave that is always there, but hidden in the www. of infinite possibilities, it is the science of Harmonics or Cycles or Proportions. Simply: Mathematics. When Pythagoras's works were retranslated, there arose a debate about the translation of the Greek word: "Logos". Many of you are familiar with the Christian expression: "In the Beginning was God" but it has been interpreted as "In the Beginning was the Word, was the Sound, was the Vibration" etc many meanings, as to what the actual word "Logos" meant.

According to Greek scholars, **Logos = Proportion**. Mathematical Proportion, as in our case or enquiry of **12:24:60**.

May this article be of some inspiration to You, at least to always weigh the evidence and to give priority to ancient Mathematical Revelation which can not lie.

Mathematics, or Logos / Proportion is the common language for the new globalizing and **fibonnacization** of the world we live in. It is time to restore it. And Mathematics is the supreme language, not French or German which, no offence, are decaying rapidly, they are limited. As much as they are precious cultures, like all cultures, we can not, on the other side of the world relate to them or understand them. But what is in common, now, for the last 10,000 years, and for the next 10,000 years to come is Mathematics. We will still have this Mathematics of **12:24:60** on different planets, on Mars and Neptune, how can it ever change, this **FIXED DESIGN**, this ETERNAL PROPORTION.

May all our children learn this Jain Mathemagics and become acquainted in the class room with these precious jewels known as:

> The Fibonacci Sequence aka The HemaChandra-Gopal Sequence
> The Pascal's Triangle aka Meru Prastera Triangle
> The Prime Numbers 4th-Dimensional Cross Symmetry of 24 Circular Numbers
> The Squaring of the Circle aka the Mystical Quadrature of the Circle
> The Vesica Piscis
> The True Value of Pi which is a revolution in Mathematics
> The 24 lengths and 24 faces of the Star Tetrahedron
> The Value of Intuitive Mathematics

Finally, I would like to discourse on the meaning of the beginning of the Fibonacci Sequence:

0, 1, 1 and the Journey to Infinity:

What does this repetition of the "1, 1" truly mean.

We know that "0" is the Void, Source, Bindu, Emptiness, yet The All. It is the crucible or womb for the possibility of any God Manifestation.

If "1" represents "God", then the occurrence of another "1" suggests that In The Beginning, "God" duplicated Him-Her-Self, to see reflected the Divinity of Self. God-Dess-Ence needed to make Him-Her-Self **SELF-SIMILAR**, thus the "1" became the "1" (not the traditional view that the "1" became the "2" which meant from Oneness we entered the binary Die-Mention of Duality, no, there was never ever any real separation as we are in all things and the true meaning of ascension IS TO SEE GOD IN ALL).

Thus God-Dess-Essence embedded Her-His **Unity-Consciousness** into all things and journeyed or quested in search of the ultimate light vehicle possible for great expansion or contraction/compression which was the Sphere in the Process of Implosion or the harnessing of the **Toroidal Domain**. (This only means that the shape of your blood platelets are doughnut-shaped!). God-Dess-Essence needed to keep multiplying without being destroyed, so HeShe or SheHe chose the most simplest of all arithmetical (adding) and geometrical (multiplying) sequences, combined them and thus the infinite process forever continued when 1+1 became 2 and the 1+2 became 3 and 2+3 became 5, and the 3 + 5 became 8, a mathematical wavelength concept known as **SHARING**. (This only means the special place where the elbow bends, or where the knee bends, or the mathematics of the belly-button). So well disguised was this Unity Consciousness of "1,1" that it could **EMBED** in the atom or the stars, in the floret of the sunflower or the pinecone, but ultimately its secret head-quarter's office is the spiral staircase that is your D.N.A that allows access to the micro and macro worlds by sensitive understanding of The Laws of Time and Space that we call Proportion or Logos and is mathematically described as 12:24:60, a Trinity of Frequencies, a timeless Pine Code that obeys the embrace of Mother Earth and Father Sky who teach their Children of the Crystal, Mineral, Star and Earth Domains to fractally embed this Knowledge in their Heart. (Though this "Fractal Heart" is only a belief. How do we integrate the fact that our native Australian Aboriginal Elders of the Dreamtime declare that the Sun, is Female!). How do we make the essential relationshift from the single Head of strict belief, to the open source of Twin Hearts.

So really, as the logical left-brained men of this world analyze and rationalize and write such articles as this, a right-brained intuitive woman could merely smell the Rose with such genuine depth and feeling that this mere act of smelling could access all Knowledge. "How beautiful the nesting and embedding of those petals in such gorgeous symmetry and chaos", she surmises. Thus the time of evolution has brought Man and Woman to become "1 and 1" or Androgynous, time for Men to learn how to Smell the Rose, and learn how to Cry, and Woman to lay the bricks and master Mathematics mentally, such that there is no distinction between the faculties of Man and Woman.

This article I pray will help heal any agonizing memories of being mathematically wounded. If you disliked mathematics as a school student, you probably did the right thing by avoiding it, as your Higher Self switched off in fear of being bombarded by incorrect conditioning, so that is good.

If you were deemed "**dyslexic**" at school, then consider yourself Gifted. Our well-intentioned teachers failed to address us visually and bombarded us with meaningless simultaneous equations and differential calculus that had no meaning for us, so we smartly switched off in the classroom, and for mere neural entertainment we began to flip, rotate, reverse and mirror-image the tirade of numerical data for sanity's sake. So congratulations if you were dubbed dyslexic or dumb, as your Higher Self's thermostat had to switch off to avoid the excessive conditioning.

But it is time now to restore the imbalance and recognize that all Mathematics is a Sacred and Universal Star Language.

It only needs now to be taught coherently, simply and visually in all schools of the world so that it switches on the remaining dormant codons of DNA igniting a child's consciousness to remember that once Mathematics was considered **a most Beautiful and Anointed subject**.

Thus in summary, we can conclude that our ancient cultures had a great understanding of the harmonics that pervade the universe. Typically, the two most important mathematical bases are **Base 9** and **Base 12**. The one and only cosmology that obviously depicts both these bases is the Sino-Tibetan calendar:

Fig 23
The Sino-Tibetan Cosmology or Calendar depicting the outer ring of 12 animal signs girdling the central motif of the Magic Square of 3x3 composed of 9 cells or the first 9 counting numbers.
Essentially, Base 12 united with Base 9 rings the 108 Bells.

So why was this calendar so important, showing a great god: **Manjushri**, incarnating as a fiery tortoise whose energy is to protect the sacred center. And why is the Lo-Shu or Magic Square of 3x3 in the center of the cosmogram, why is it not just a small intellectual motif in a corner somewhere? I believe that the combination of both the 9 numbers of the Lo-Shu and the ring of 12 animal signs, act like a sophisticated padlock or combination key ringing to the sound of 9 x 12 which equals this magic number 108 so far discussed. Does such a calendar place importance upon Time, in the sense that it refers to a specific angle in the sky when and where you were born astrologically, and if so, is this eternal recursion of the 108 harmonic of the Phi Code really a **Stargate**?
Is this Phi Code really a **TIME CODE**?

Many questions abound from these simple observations as found in Nature and Mathematics. Why would a whole culture, billions of Hindu people who acknowledge the Vedas of the ancient Indian tradition, mindlessly worship the number sri or holy108 and not even know why they are doing so? How many thousands of years has this secret been whispered?

And without these mathematical sutras or links, there appears to be something **missing** in our understanding of the whol- :

Th- conclusion that I am making h-r- is that a lot of th-s- qu-stions could not b- ask-d had it not b--n for th- ability to und-rstand th- most simpl- V-dic Sutra call-d Digital Sums or Digital Compr-ssion, bas-d on th- simple arithm-tical fact that 34 = 3 + 4 = 7, or multiplying by 11 has a uniqu- us- of this sutra, that 34x11 = 3 / 3+4 / 4 = 374 (by m-r-ly adding th- sum of th- digits 3 & 4 and sandwiching this sum in th- middl- to giv- an instant answ-r, this x-raying of imm-ns- data b-ing r-duc-d to fundam-ntal harmonics is th- subj-ct of d--p appr-ciation h-r-, that our childr-n n--d to r-m-mb-r that Math-matics is a supr-m- and Univ-rsal Star Languag- and it can be dir-ctly acc-ss-d by sm-lling th- flor-t of a sunflow-r or th- Ros-.

R-gards Jain.

PART 4

"Lost Secrets of the Phi Code" continued

The BINARY CODE Versus The PHI CODE

I am discoursing and teaching a lot on the distinction between the two most important codes:

The Binary Code Versus **The Phi Code.**

Ultimately, I am saying that we are neither this or neither that, we are both, and this can be proven geometrically and dramatically in one beautiful diagram that shows how the concentric rings of the binary code as they expand from one to two to four etc have a phi ratioed connection. Its beautiful, and teaches that we are all Codes or Sequences.

BINARY CODE: 1 - 2 - 4 - 16 - 32 - 64 - 128 - 256 - 512 etc (we were this binary code, from the original cell in our CosmoGenesis: father-mother cell was The One that became the Two that became the 4 (whose 4 centres make the Tetrahedron) that became the 8 (whose 8 centres became the Star Tetrahedron or Cube) whose final mitotic division of endless halving collapsed its superb geometry at 512 cell division and became The Tube Torus Doughnut! Yes we were once this ring shape having two holes, one for the mouth that received the universe, and one for the anus that released the universe.....and from this primal shape we unfurled into a fern-like entity whose overall shape was the golden mean spiral up to the point now where every part of our being, like where the elbow bends, is in the divine phi 1:1.618 proportion....

This is binary code: when the One became the Two, which symbolizes separation from God...... and is about Technology, the Machine or borg-hive consciousness, Electronics, the "Zero and One" of computer digital language. It about midpoints, halving, point 5 ratio or halving of the whole... Its symbol is the Equal Spin of the Sahasrara or Crown Chakra, where 1,000 spirals spin clockwise superimposed by 1,000 spirals spinning counter-clockwise. Whatever number of spirals (I suspect it is 1,008), it is balanced and equal in its counter-rotating fields.

Of note here, we can state that Nature does not do this, (Nature does 21:34 as in the Sunflower Map).

It is my observation here that the masters of ancient times who encoded this Equal Spin for the Crown Chakra, did it as a deliberate ploy to inhibit Time Travel, to create a ceiling, a barrier into the unlimited nature of the Phi Code

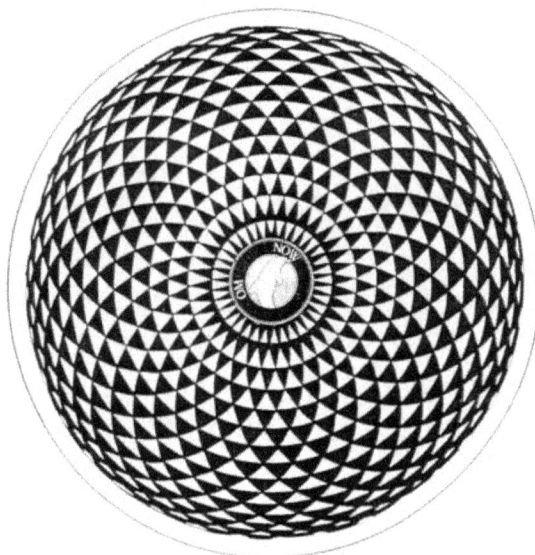

Fig 24
This is not Nature, this is Equal Spin.
Do not put this on your head!

This diagram shows what we call "Equal Spin" 36 spirals going one way, and 36 going the other way. What this does, when it is visualized at the top or crown chakra is to effectively start the motor, it merely turns it on, agitates energy, gets it spinning, but it does not go anywhere, it is like a car whose motor is turned on, but it is just idling. It does not travel. Whereas, in the next section, you will see what nature does, the 21:34 spin code which permits travel through the dimensions. In summary, Binary Code is efficient, but it is unlimited. Phi Code is Nature's choice, it is unlimited.

(For more insight, you can refer to the rare diagram on pages 99 to 101 in The Book of Phi, Volume 3, published 2010, on the Cosmic Vibrations of the Powers of Two, the Masonic chart of the wavelengths of Colour, sound, heat, gamma rays etc... noticing how quickly this Doubling Sequence can achieve high numbers).

THE ONE BECOMES THE ONE

Phi Code:
0 - 1 - 1 - 2 - 3 - 5 - 8 - 13 - 21 - 34 - 55 - 89 - 144 etc

This is where the One becomes the One, not the One becomes the Two, and is expressed by the first two numbers: "1 - 1".

One becomes the One, is symbolic of the Creator creating another Self to view the Creation of Self as it expands out in Life.

The Fibonacci Sequence, as it is incorrectly know, is really a **Trinary** Sequence: where three components are considered:

Past + the Present = or Creates the Future:

0 + 1 = 1
1 + 1 = 2
1 + 2 = 3
2 + 3 = 5
3 + 5 = 8

In my next books, THE BOOK OF PHI volumes 5 & 6, available in 2013, you will learn that there are 3 Phi Codes or 3 Dials of infinitely repeating and digitally compressed 24 digits, all of which sum to 108.
(see Index of digitally Compressed Sequences at the end of this book).
This means that if you take any two random numbers in the universe, say the numbers 3 and 8, and keep adding them in the trinary fashion, or Fibonacci-like style, where the next term becomes 3+8=11, and the next terms becomes 8+11=19 etc and compress the digits to single digits, you will always come back to one of the 3 Phi Dials. Essentially, when we dissect the Phi Code and its related Fibonacci Sequence, its skeletal structure is based on 3ness, a veritable Trinitization of how the Invisible World works.

PHI RATIO SHOWN GEOMETRICALLY IN THE BINARY CODE:

(taken from: http://www.phinest.com/)

The following diagram and its proof, is jaw dropping evidence that the Phi Ratio is basically in everything.

Up till now, we thought that the opposite of the Phi Ratio, is midpoints of the Binary or Doubling Sequence, but now, there exists a remarkable mathematical proof that the Phi and Binary Codes are married.

Concentric Circle Construction:

Here's a construction using 3 concentric circles whose radiuses are in a ratio of 1:2:4 which is really a geometric form of the binary code.

Method:

Draw a tangent from the small circle through the other two, crossing points A and B and extending to G.

The ratio of the length of segment AG to segment AB is Phi, or 1.6180339887...

Proof:

(the symbols: "*" means to multiply, and the exponent " $3^{1/2}$" means "the square root of 3 or $\sqrt{3}$)

$AB = 2 * 3^{1/2}$ and $AG = 15^{1/2} + 3^{1/2}$, which by factoring out the $3^{1/2}$ can be reduced to a ratio of 2 to $(5^{1/2} + 1)$, or Phi.

This same ratio is expressed as: $(1 + \sqrt{5}) \div 2$

Since $\sqrt{5} = 2.236067978$ then

$(1 + \sqrt{5}) \div 2 = 3.236067978 \div 2$

$\qquad\qquad\qquad = 1.618033988...$

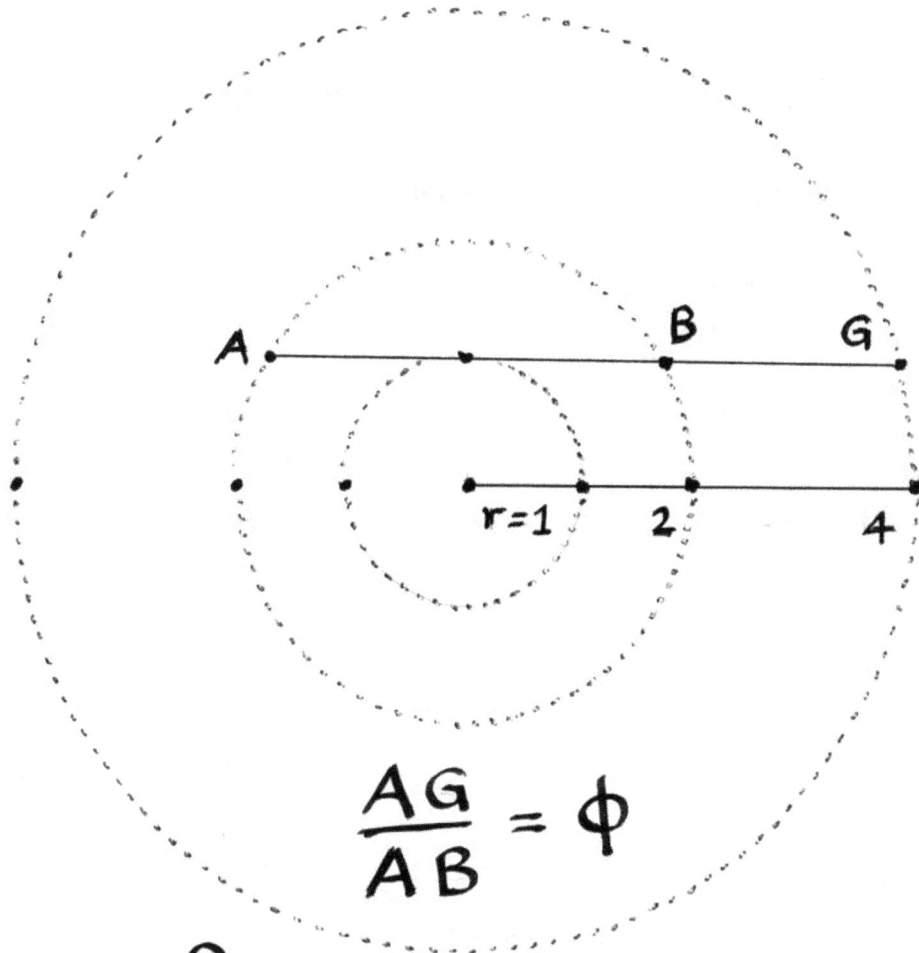

$$\frac{AG}{AB} = \phi$$

Geometric Proof that Phi exists in the Binary Code.

Fig 25
Geometric Proof by Sam Kutler
proving that Phi Ratios exist in the Binary Code
when expressed as concentric rings that keep doubling.

(this construction was developed by Sam Kutler and submitted to
www.phinest.com by Steve Lautizar)

This amazing diagram basically proves that the Phi Ratio is in everything, even
the dreaded binary code that some scholars relate to the maths of technology,
aliens, borg-hive-mind and computer chips!

GEOMETRIC PROOF for the BINARY / PHI CONNECTION:

Construct 3 concentric circles that represent the doubling sequence of 1:2:4
Let "O" be the Origin or Centre.
With Radius = 1 unit, make a circle.
With Radius = 2 unit, make a circle.
With Radius = 4 unit, make a circle.
Draw a vertical from "O" to touch the Unit Circle at "C".
At this point of C draw a long horizontal tangent that touches the second circle at A and B, top left and top right respectively, and touches the third circle at "G".

$OA = OB = 2$
$CB = AC = sq.root(2^2 - 1^2) = sq.root(3)$
$AB = double\ this\ value = 2sq.root(3)$
Find CG
$OG = 4$
$CG = sq.root(4^2 - 1^2) = sq.root(15)$
$AG = AC + CG = sq.root(3) + sq.root(15)$

Examining the Proportion of the whole length AG to AB
$AG / AB = sq.root(3) + sq.root(15) ÷ 2sq.root(3)$
Dividing top and bottom by sq.root(3)
$= sq.root(3)[1 + sq.root(5)] ÷ sq.root(3)x2$
$= [1 + sq.root(5)] ÷ 2$
$= 1.618033988749884$
$= \phi$

Chart for Jain's ideal of GLOBALIZATION

BINARY	PHI CODE
Binary Doubling Sequence	Trinary Sequence
1 - 2 - 4 - 8 - 16 - 32 - 64	1 - 1 - 2 - 3 - 5 - 8 - 13 - 21
The One Becomes Two	The One Becomes One
As in Human Cell Division (Mitosis)	As in The Plan of Plants and Planets, Human Canon
Cosmo-Genesis	All bio-Proteins are Pent
Symbol = 2^n Two to the Power of 'n'	Symbol = Phi = ϕ Adoration of the Pentacle
Limited	Unlimited
Mindless Storage of Data	EnGrailed: is SomeOne who Understands this Universality of the Fibonacci Sequence.
Energy of the Technological / Machine	Energy of the Etherical / Spiritual
External MerKaBah	Inner MerKaBah
Age of the Computer	Age of the Third Eye Awakening
Mathematics on the Electronic Calculator	Mathematics performed in The Inner Mental Screen
Fear	Love
Control and Separation	One World: Globalization
Fractionation	Fractal
Political Symbol = Pyramid of Mutli-Level-Marketing (MLM) Hierarchy, greed, plutocracy, deception	Political Symbol = Torus: allows no Hierarchy! Equality of all Professions, Holographic Unity
Many Religions	One Religion
Symbol = Equal Spin as on Sahasara Crown Chakra: Turns motor on, but no Travel,	Symbol = Nature's Sunflower Code of Phi-Ratioed Spin 21:34 Permits Time Travel from Macro to Micro
Secret Initiations	No Secrecy
Indoctrination	No Doctrines
Morales	So What
Regulations	All is Possible, Permissible
Factory-Style School Maths	Vedic and Jain Mathemagics
Control	Fibonaccization of the Human Species

Fig 26

CHART for Jain's ideal of GLOBALIZATION
Or SPIRITUAL COMMUNISM

One State
One Human Society
One Monetary System
One Train Track Gauge
One Mathematics
One Language
One Love
One

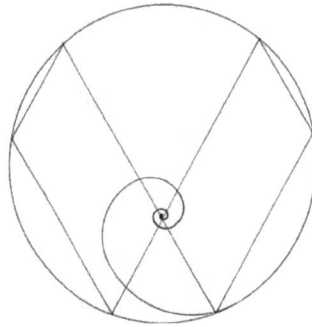

Binary (Techno) Versus Phi (Biological)

Awaken from the Dream of Duality

There is no Versus

Binary and Phi are One

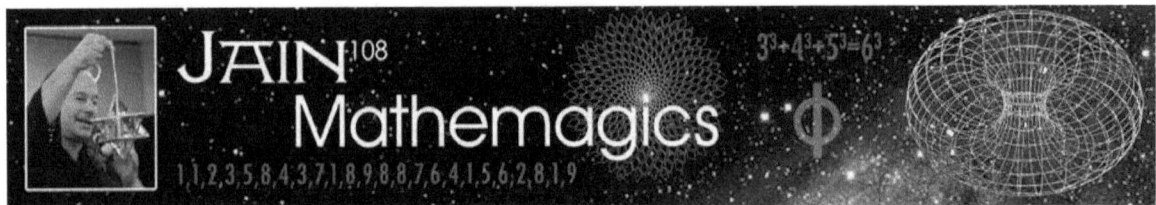

11-11-2011

As a follow up to this series of articles, Jain will be giving a one day seminar that will illuminate this body of knowledge. Entitled: **"The Vedic Mystery of 108 Revealed"** on the 11th November, 2011, in Byron Bay. Since his return from India, Jain has been hailed as the first western scholar to explain to the eastern world the deeper meaning of the holy Sri 108.
The Indians are now calling him by the name of Jain 108.

1 – THE BOOK OF PHI, Vol 1 Subtitled: The Living Mathematics of Nature. Book $60. eBook $44, 172 pages	**2 – THE BOOK OF PHI, Vol 2** Subtitled: **In The Next Dimension.** Book $60. eBook $44, 188 pages
3 – THE BOOK OF PHI, Vol 3 Subtitled: **The 108 Codes,** an Introduction Book $65. eBook $44, 216 pages	**4 – THE BOOK OF PHI, Vol 4** Subtitled: The 108 Codes, The Linear Phi Code 1. Book $65. eBook $44, 197 pages

JAIN'S 4 BOOKS ON PHI

Email Contact: **jain@jainmathemagics.com**
For more detailed information on the Lost Secrets of the Phi Code,
You can order from Jain's website, 4 highly informative books and ebooks.

5 DVD SET, Series
– THE LIVING MATHEMATICS OF NATURE,

Part 1 of 5: INTRODUCTION. 2 Hours. $45
And the complete **5 DVD SET** 10 Hours. $200.

This Dvd will give the reader a total overview of Jain's Life's-Works covering the 4 distinct topics that he teaches world-wide:
The 5 Dvd Set
1- Introduction, being a summary of the following 4 titles
2- Vedic Mathematics aka Rapid Mental Calculation
3- Magic Squares
4- The Divine Phi Proportion, and
5- Three-Dimensional Geometry of the 5 Platonic Solids

Distributed by Jain Mathemagics
The Living Mathematics Of Nature
5 DISC DVD SERIES

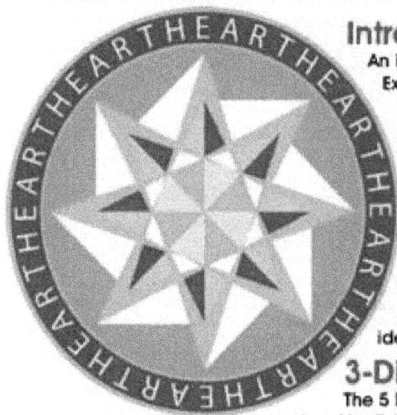

Introduction to Jain Mathemagics
An introduction to the 4 topics below.
Explains each topic, providing good overview of work.

Vedic Mathematics
Rapid Mental Calculation.
No more calculators. Increases your memory power!

Magic Squares
Translating numbers into Atomic Art.
Teaches children pattern recognition.

The Divine Phi Proportion
Explores the Geometry of Flowers
identical to the Human Canon!

3-Dimensional Geometry
The 5 Platonic Solids
adored by Pythagoras and his community.

Videos & Books Available Online...
www.jainmathemagics.com

Distributed by JAIN MATHEMAGICS
777 Left Bank Road Mullumbimby
NSW 2482 Australia

Ph: (02) 6684 4409
Email: jain@jainmathemagics.com
www.jainmathemagics.com

New DVD Series

Copyright © Jain Mathemagics 2005

Testimonial:
"Sunday was so exciting, the revelation you receive for the recursive Phi pattern is at once so staggering... yet excitingly familiar, my physical perception actually shifted. The potential of your rediscovery leaves me breathless. Thank you for bringing the joy of this pure science into our lives".
Ingrid Burke, Sunshine Coast, Australia.

~ CHAPTER 5 ~

PRIME NUMBERS
4TH DIMENSIONAL TEMPLAR CROSS

CHAPTER CONTENTS:

PART 1

Definition of Primes and Ulam's Rose

A prime Number can not be divided by any other number except by 1 or by itself. They are like the Atoms of Creation.

For 2,000 years, we have been told that no distinct pattern or symmetry exists within this infinite nonsense sequence!

Prof. James McCanney (and **Reginald Brooks**) recently discovered waves of symmetry in the Prime Number Sequence, which means that all mathematical books need to be revised. Up till now, the military and internet encryption systems were based on the largest Prime Numbers known to us, but since this amazing discovery, NASA are trying to shut down this revelation. One of my goals is to train teachers to learn quickly how to determine the next Prime Numbers and understand the inherent symmetry and patterning in this code.

Of all numbers, the Primes are of royalty. You will see in a short while why the Queen of England wears the secret insignia for Prime Numbers which is the Prime Number Cross, what you may know of as the Maltese Cross. See Fig 1. One must wonder why this symbol has been so special over the centuries, and why it was worn on the Heart/Solar Plexus Chakra.

Currently the world daftly thinks there is no pattern in this sequence. Top professors still play around with notions like:

Why do prime numbers occur at such inconsistent intervals? Could there be one single formula that predicts all prime numbers, or something that can be said that is true for all prime numbers? Is there any kind of regularity in the appearance of primes?

Of course "They" know there is a pattern, they just don't want you to know it, that is why all the mathematics books are disharmonic or in error, and why children shut down disappointed that their insipid factory style of maths has been watered down and is no fun at all.

The Order of St. John
(A SHORT HISTORY)

Fig 1
**The Prime Number Cross (Maltese Cross)
worn by members of the Order Of Saint John**
circa the 11th and 12th Centuries in the Holy Land.

HOW TO DETERMINE THE PRIME NUMBER SEQUENCE
USING THE SIEVE OF ERASTOSTHENES:

Here is a simple and ancient technique to derive the Prime Numbers from the first 100 naturally consecutive numbers.

In the space below, mark or shade in or circle all the numbers that are prime, by systemically ruling out all the multiples of 2, 3, 4, 5, 6, 7, 8 and 9.

Observe that ever 2nd column is comprised of only even numbers, so you can systematically draw a line through the centre, as if to cross them out.

1	2	3	4	5	6	7	8	9	10
11	12	13	14	15	16	17	18	19	20
21	22	23	24	25	26	27	28	29	30
31	32	33	34	35	36	37	38	39	40
41	42	43	44	45	46	47	48	49	50
51	52	53	54	55	56	57	58	59	60
61	62	63	64	65	66	67	68	69	70
71	72	73	74	75	76	77	78	79	80
81	82	83	84	85	86	87	88	89	90
91	92	93	94	95	96	97	98	99	100

Fig 2

The first100 counting Numbers in a 10x10 array.

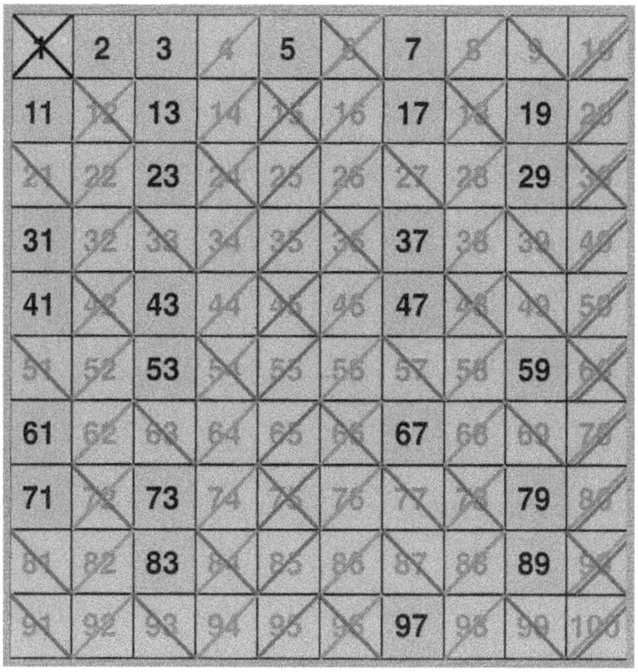

Fig 3

The Sieve of Eratosthenes created by an ancient Greek mathematician.
The numbers from 1 to 100 are crossed out in certain ways
until only the indivisible Primes remain.

The Sieve of Eratosthenes:

Eratosthenes of Cyrene (c. 276 BC – c. 195 BC) was a Greek mathematician, elegiac poet, athlete, geographer, and astronomer. He made several discoveries and inventions including a system of latitude and longitude. He was the first Greek to calculate the circumference of the Earth (with remarkable accuracy), and the tilt of the earth's axis (also with remarkable accuracy).

In mathematics, the **Sieve of Eratosthenes** is a simple, ancient algorithm for finding all prime numbers up to a specified integer.

Consider a contiguous list of numbers from two to some maximum number like 100.

Strike off all multiples of 2 greater than 2 from the list.

The next lowest, uncrossed off number in the list is a prime number.

Strike off all multiples of 3, then 4 then 5 etc until all the numbers remaining in the list are prime.

It's a good idea to learn these primes, as if they were a cosmic phone number to God.

Fig 3a
The list of Prime Numbers from 1 to 50

ULAM'S ROSE

Before we show how the Prime Number Cross is extruded out of the 24-ness obedient to Phi's mystery patterns, here is yet another famous pattern found in the so called non-sense symmetry of the prime number sequence. This one is dedicated to Stanislav Ulam.

It is important, because you will clearly see yet again, how numbers are turned into Art.

Was there a pattern in the Prime Number Sequence?

No one could supply the world with an answer and primes were believed to occur randomly. The excitement about primes flared up even more in the wake of boredom of a devoted 20th Century math-magician named Stanislav Ulam.

He put down the number 1 as the bright shining center of a universe of numbers that Big Banged outwardly in a spiral.

Fig 4

STANISLAV ULAM who found an organic form in the infinite web of Prime Numbers, now known as Ulam's Rose.

73	74	75	76	77	78	79	80	81
72	43	44	45	46	47	48	49	50
71	42	21	22	23	24	25	26	51
70	41	20	7	8	9	10	27	52
69	40	19	6	1	2	11	28	53
68	39	18	5	4	3	12	29	54
67	38	17	16	15	14	13	30	55
66	37	36	35	34	33	32	31	56
65	64	63	62	61	60	59	58	57

Fig 5

**Ulam's Spiral of Numbers beginning from the center 1
and circling this forever clockwise.**

To see this concept of spiralling numbers issuing forth form a divine centre, here is a simple spiral path that you will recognize, almost like a labyrinth path, a journey from the centre then outwards, and vice-versa.

Fig 5a

Ulam's Spiral seen as a spiral pathway

This is what he thought: What would happen if we were to write the natural counting sequence of numbers, that is the consecutive numbers as in 1 – 2 – 3 – 4 – 5 – 6 - etc upon graph paper or gridded paper, in a spiral fashion, starting from the number 1 in the centre, and radiating outwards to infinity?

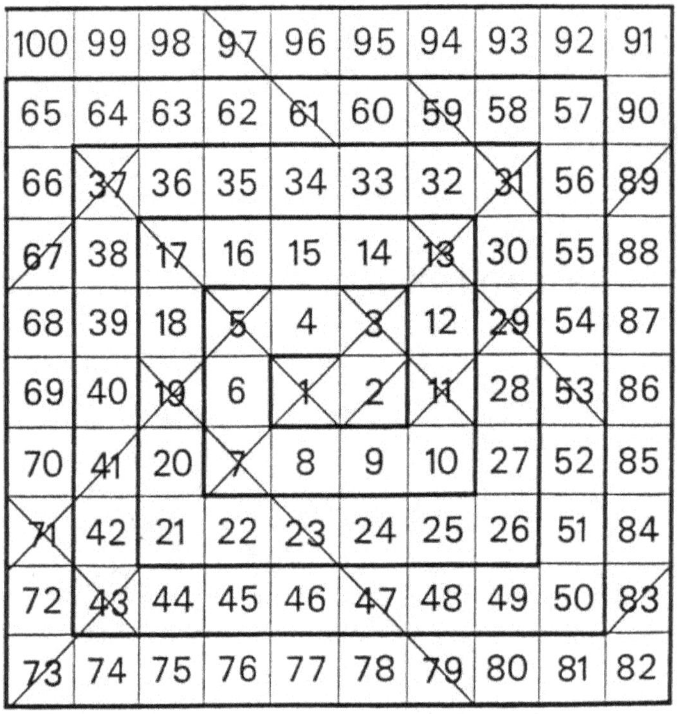

Fig 5b

Ulam's Spiral using numbers from 1 to 100, seen as a spiral pathway
marking all the Prime Numbers with a Cross.

Then, since we are mathematical explorers here, or pattern hunters, we begin to mark in with a cross or colour in all the known prime numbers. Would we expect to see a pattern, as we approach infinity?

You can see in Fig 5b that already there is a sense of a pattern emerging, enough to want to continue with larger numbers.
Much to his amazement the prime numbers appeared to gravitate towards diagonal lines emanating from the central, see Fig 5c.

Yet there was no apparent rule that forced *all* prime numbers upon a diagonal line like that. Most of them sat on or in the vicinity of a diagonal, but some obviously didn't. Ulam ran home and expanded the spiral to cover a much larger portion of the number sequence. The strange pattern persisted. Primes had a tendency to occur in clusters and all clusters tended to make a beautiful image that could not be predicted.

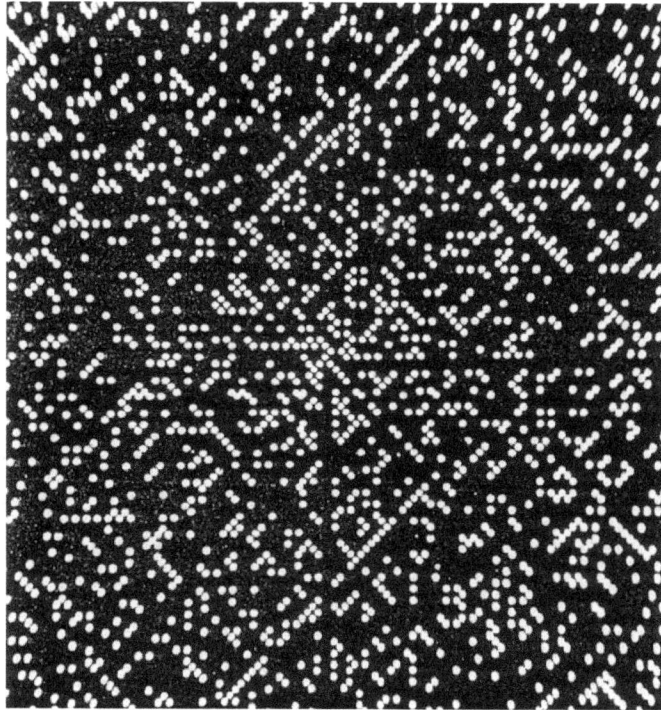

Fig 5c
Ulam's Spiral using numbers from 1 to 10,000.

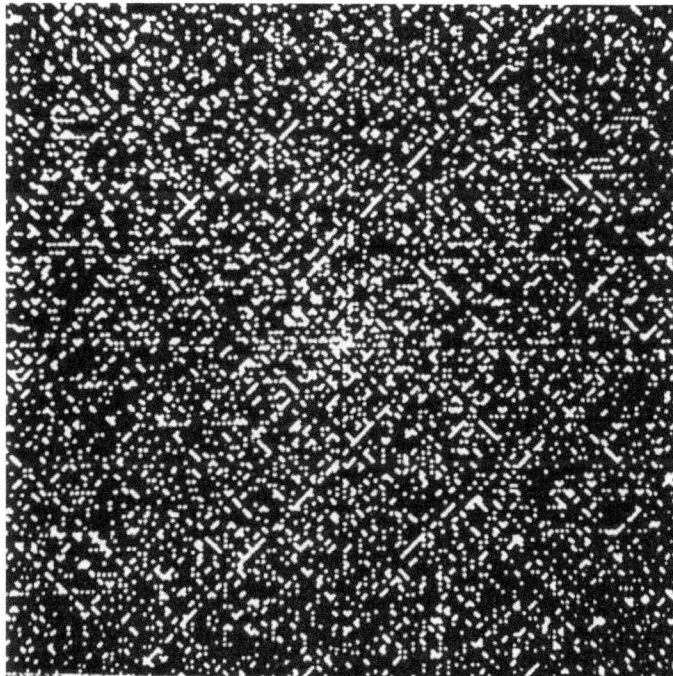

Fig 5d
Ulam's Spiral using numbers from 1 to 65,000.

The pattern above could look like the city night lights viewed from an aeroplane; there is a sense of order amidst the disorder.

What would it look like if we plotted all the primes into the hundreds of thousands? See Fig 5e.
It looks like something out of nature but in fact it's the prime numbers from 1 to 262,144. Like water molecules huddled together to make a snow flake according to some basic design, prime numbers huddle together to make the Ulam Rose.

The **arythmophilic** world responded in awe. There's true, unpredictable randomness in the prime number sequence! Numbers are as beautiful as nature! And up to this day every book on popular mathematics uses the word *random* in direct relation to prime number distribution.

Again, Ulam's Rose is another feather in Sri Technology's hat, for without the power of the silicon chip, we could not have had the computers to extract this amazing pattern out of the void of infinity.

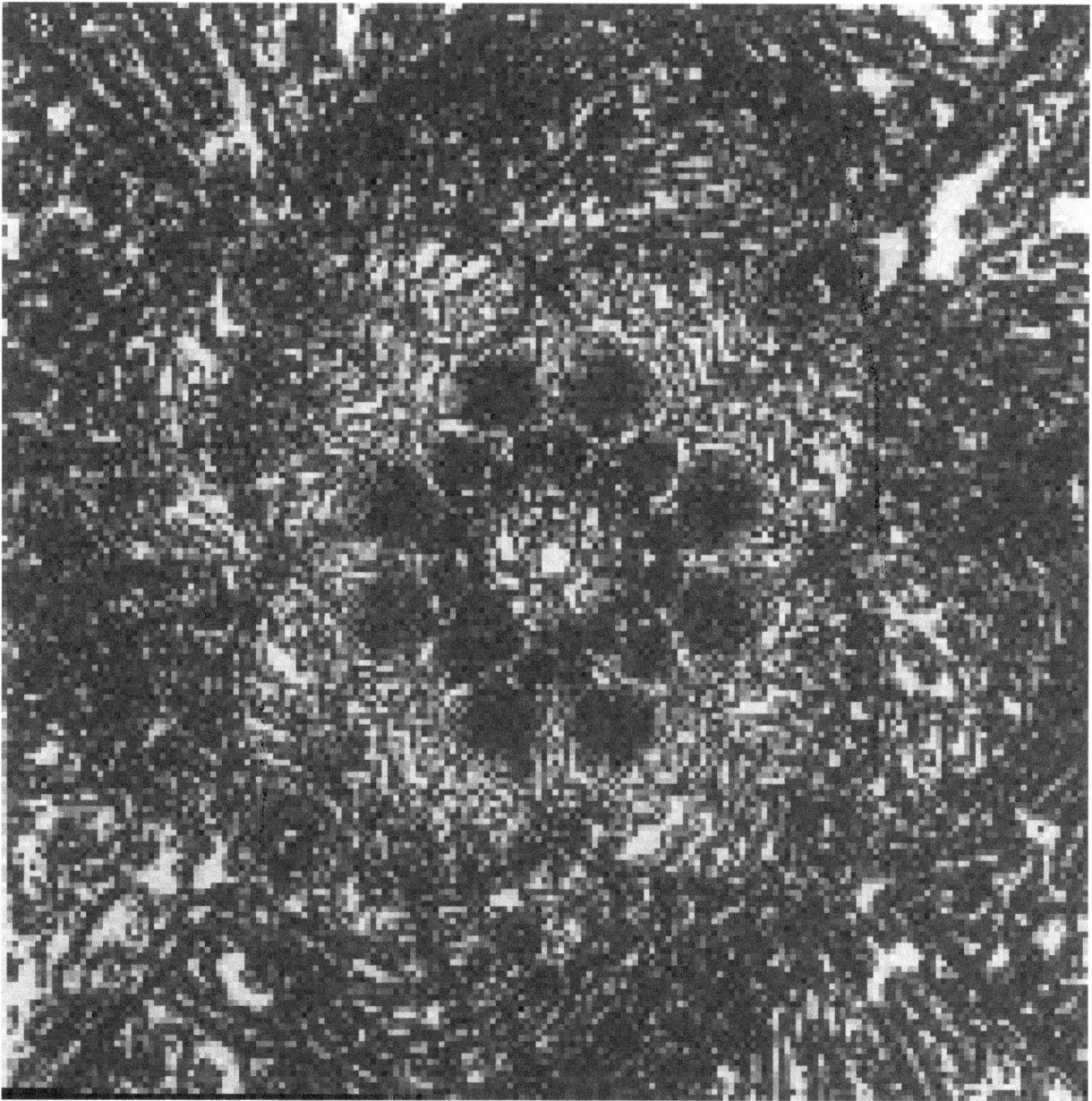

Fig 5e
**Ulam's Rose of Prime Numbers:
is not a random pattern but is a flower!**

[Credits: The Ulam Rose of 1 => 262,144 used here is an embellishment of an image originally created by Jean-François Colonna ©1996, CNET and the École Polytechnique, Paris France. The picture used here comes out of *Cracking the Bible Code* by Jeffrey Satinover, M.D.]

Fig 5f
The True Rose,
Ultimate symbol of Sacred Gaiaometry
that demonstrates Perfect Embedding,
Recursive Divine Proportioned Enfolding,
Self-Organized and Shareable
Non-Destructive and Forever
the art of how to Get Fractal
the art to feel the Tingle of your Fingertip
the Living Curvation
the Mathematics of Beauty
the Bliss in a Child's Eye.

Here is a bit more theory on how Prime Numbers arrange themselves. Notice the 6 columns!

The Distribution of Primes in columns of 6 $(6n + 1)$ or $(6n - 1)$					
1	2	3	4	5	6
7	8	9	10	11	12
13	14	15	16	17	18
19	20	21	22	23	24
25	26	27	28	29	30
31	32	33	34	35	36
37	38	39	40	41	42
43	44	45	46	47	48
49	50	51	52	53	54
55	56	57	58	59	60
61	62	63	64	65	66
67	68	69	70	71	72
73	74	75	76	77	78
79	80	81	82	83	84
85	86	87	88	89	90
91	92	93	94	95	96
97	98	99	100	101	102
103	104	105	106	107	108

Fig 6

The Distribution of Primes in columns of 6 $(6n + 1)$ or $(6n - 1)$

This table is well known to most mathematicians.

All prime numbers have been shaded in, showing a preponderance in the 1st and 5th columns.

They recognize these 6 columns that harbour the prime numbers, but because they could not think circularly, as in the **Wheel of 24**, they could not grok or perceive the immense symmetry. Symmetry is the charge, is what we are seeking, and therefore teaching.

PART 2

PRIME NUMBER CROSS GENERATED From CONCENTRIC RINGS Of 24

Why have we been told for 2,000 years that there is no distinct patterning to the infinite sequence of Prime Numbers.

Yet observe what happens when you write the natural counting numbers in concentric rings of 24, suddenly there opens up a 4th Dimensional Cross, known by the ancient Egyptians and used by the Knights Templars who wore this on their breastplate or heart chakra.

Why is this number 24 important in unlocking the keys to Time Travel (24 hours to the day!) or the Physics of Time Bending.

▲ *Four salmon tails in the arms of Simo in Lapland form a cross for the see of Uppsala.*

Fig 7
(4 Salmon Tails, arranged into a Maltese Cross symbol, Lapland)

Write all the numbers from 24 x 4 = 96 in the diagram below of concentric circles displaying Wheels of 24. The direction of writing is clockwise. Set 1 starts at the centre point (1) and goes to 24, after which there is a jump onto the next concentric ring which begins Set 2 for the numbers from 25 to 48.

Set 3 goes from 49 to 72, and Set 4 goes from 73 to 96.

Then carefully circle all the numbers that you know are Prime.

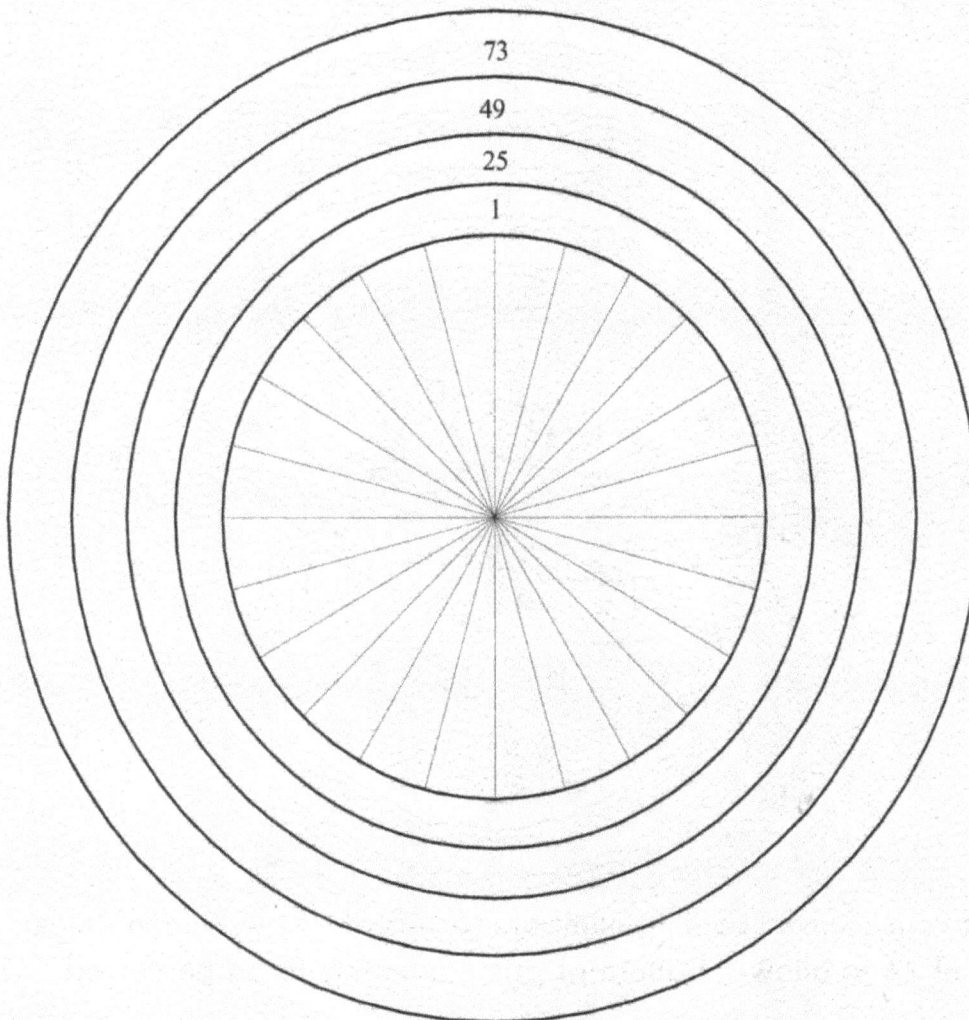

Fig 8a

Worksheet with concentric rings of 24 divisions.
The first 96 consecutive counting numbers (4 sets of 24) written in Rings
or Wheels of 24 to allow a pattern of primes to be perceived.

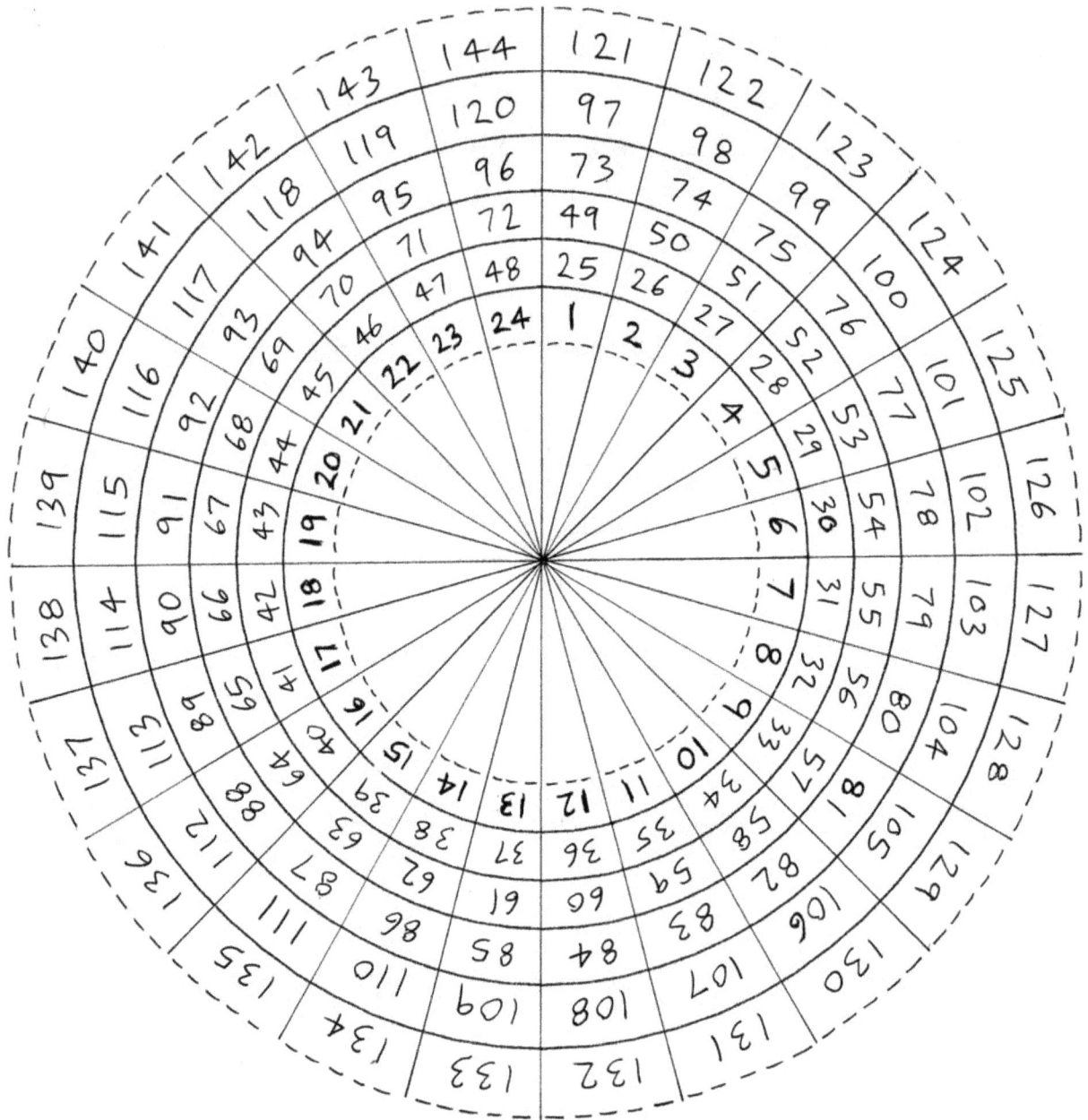

Fig 8a Solution
The first 144 consecutive counting numbers (6 sets of 24) written in Rings or Wheels of 24 to allow a pattern of prime numbers to be perceived.

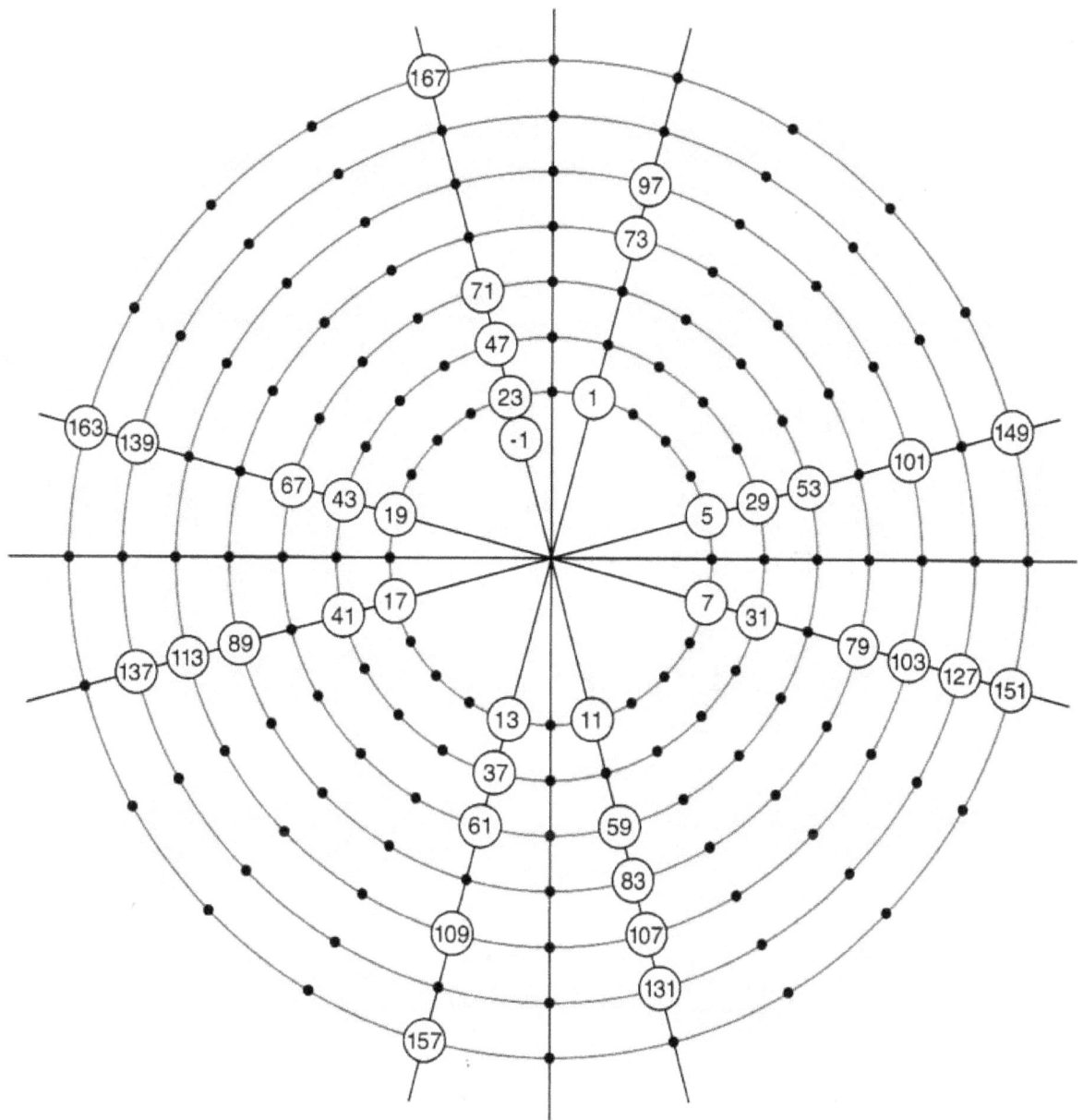

Fig 8b
All the Prime Numbers have been circled
in the first 7 concentric rings of 24 divisions.

Here is another version of the same diagram, from the work of Peter Plichta a German bio-chemist who claims that this was ancient Egyptian knowledge.

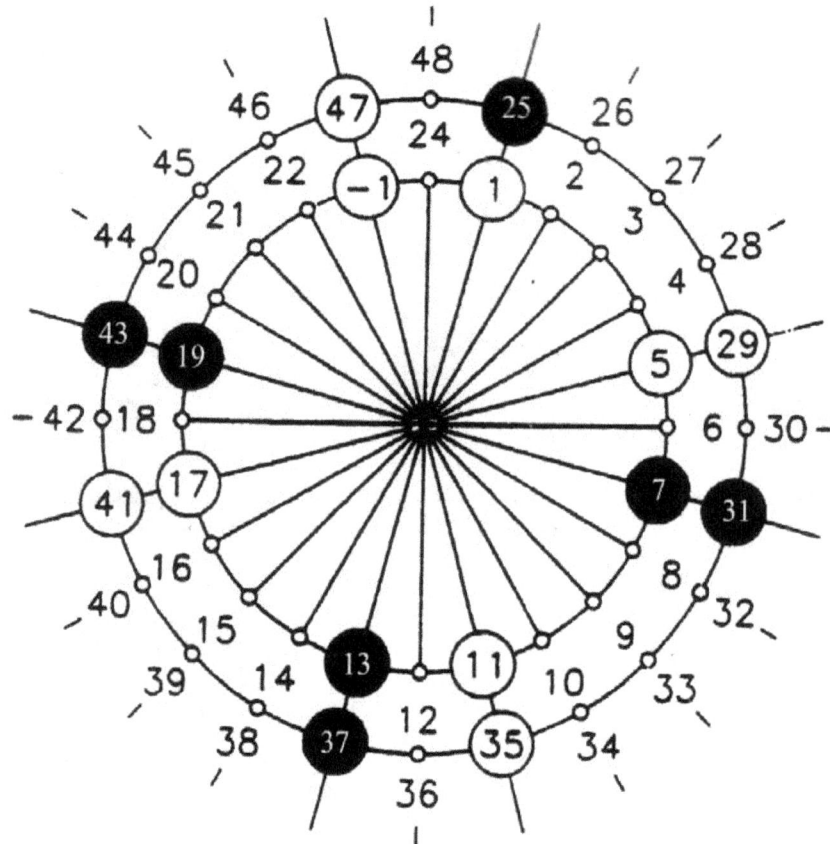

Fig 9

Peter Plichta's version of the Prime Number Cross

Did you notice that all the Prime Numbers appear to be aligning themselves along 4 distinct axes or compass points?

This could only happen by dividing the circle into 24, not any other number. The clue was given before in Fig 6 where I tabled The Distribution of Primes in 6 columns of (6n + 1) or (6n − 1). Since 6 is a factor of 24, it raised 24-ness to being the grand secret to open up this mystery.

In the next diagram I will connect these 4 diagonals or axes where the Prime Numbers lie, and investigate the pattern that emerges:

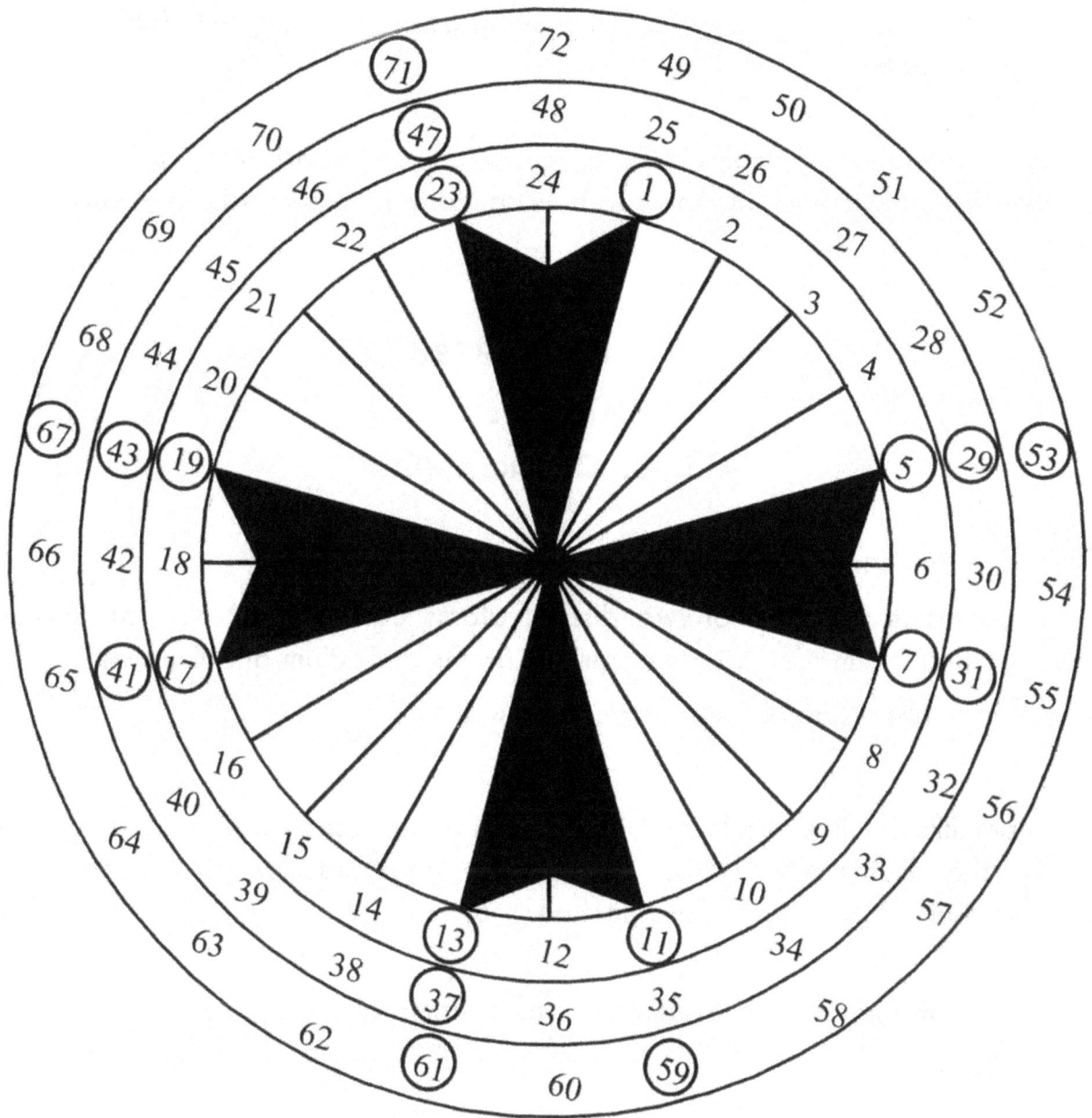

Fig 10

The 4ᵗʰ Dimensional Templar Cross, highlighted.

Circular Solution to the Prime Numbers on the concentric rings or
Wheels of 24. Here there are 4 sets of 24 consecutive or sequential numbers
(ie. Natural counting numbers 1,2,3,4,5,6 etc) making a total of 96 numbers.
All the Prime Numbers are circled.

The secret was to apply the concept of **24ness** which is the key that also opens up the 108 Codes based on hidden pattern in the Divine Proportion and Fibonacci Numbers. This Key of 24 is really a Base 12 Code, like Base 9, it has a galactic signature.

This 4th Dimensional Templar's Cross, worn on the Heart was based on the division of the circle into 24. This shape is called the IcosaTetraGon = having 24 sides.

PRIME NUMBERS
ARE THE
ATOMS OF CREATION

Atoms can be connected to other atoms to make compounds.

A compound is make up of two distinct atoms eg: Salt is $NaCl_2$ or Sodium Chloride, and it is made up of two various atoms of Sodium and Chorine.

Similarly, numbers can be atoms or compounds.

A number like 24 is divisible by other numbers, its factors are 1, 2, 3, 4, 6, 8 and 12, therefore it is not prime. It is not an atom of creation, as it is formed or compounded by the multiplication of 2x12 or 3x8 or 4x6.

All numbers though can be written in terms of Prime Numbers, eg:

39 = 3x13.

Shown here are 4 axes of prime digital information. A line is 1-Dimensional, a flat square is 2-Dimensional whose diagonal makes the "x" shape, a cube of 3 axes is 3-Dimensional, and thus 4 axes must form a 4-Dimensional Cross:

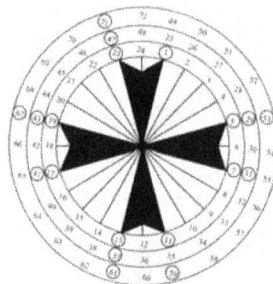

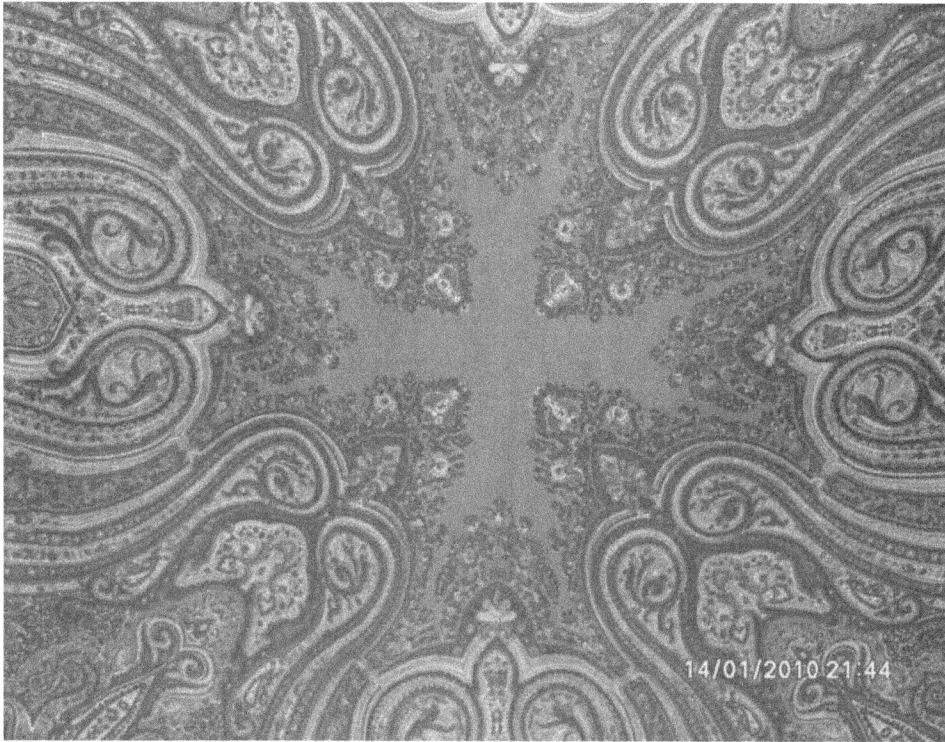

Fig 11

Two patterned paisley versions on fabric that resemble the Templar Cross!
(photos by Jain)

Below, the same cross, but highlighting the original colour coding in red, but shown below in black

Historians speculate about the choice of red colouring, perhaps for the blood that was lost when the Order of St John, Crusaders and other Masonic-like groups were persecuted for this knowledge around the 11th and 12th Centuries.

Author Kathleen McGowan's research towards the lost red symbol to be ascribed to Mary Magdalene's blood-line. (some of her books alluding to this can be read, in this order: "The Expected One", "The Book Of Love" and "The Poet Prince").

Worldwide, we know that the current symbol for rescue emergency healing is the Maltese Cross form also known as St John's Cross.

This is shown in Fig 12 on an ambulance vehicle here in Mullumbimby, Australia.

Fig 12
**Red Cross, symbol Emergency or Rescue Healing,
shown on an ambulance vehicle.**

(I am pictured next to this symbol as a "Thank You" to the servicemen who saved my life in Blackheath, west of Sydney near Katoomba in 1984 when a hired thug lunged a blade through my heart chakra. I survived, to ultimately write this book.

Spending 2 days out of my body gifted me with more insights into these da-Vinci-like Codes:

the secret is the ability to translate Number into Art).

Only Nature's intelligent choice of 24-ness generates the divine symmetry of the Phi Code 108 Mysteries.

This symbol is found even on small coconut islands out in the South Pacific:

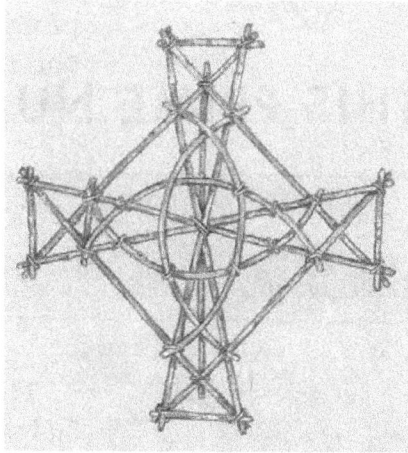

Fig 13

Prime Number Cross Form seen in Mattang from Polynesia

Where else have you seen this ancient hidden symbol?

THE TEMPLAR CROSS, a very sacred image worn on the Templar's heart. (They kept the bloodline of Jesus and Mary Magdalene alive, whose children started the dynasties of France, Italy and in England, the King Arthur legends. The Christian Church at the time tried to wipe out the memory of these upholders of the truth who dared to wear the Knights Templar Cross where nearly a million "Cathars" were genocided in the Holy Wars).

In contemporary times, its interesting to ask why the Queen of England wears the Prime Number Cross on her cloak?

She and her ancestry have claimed this supreme mathematics. It is perhaps one reason why the English Empire has risen so high. That is why Prime Numbers are not taught thoroughly at school, and it is taught that there exists no distinct pattern in prime numbers, that they are a nonsense sequence, so that you do tread in that direction, in fear that you may become awakened or **ENGRAILED** by realizing the truth that there exists a divine order in the universe.

Have a guess who wears this the 4th Dimensional Templar Cross! This ancient Knowledge of 24-ness is partly why we have 24 hours of the day etc and a Base 12 Imperial system of Measurement. Whoever has this secret, royal and confidential knowledge has the potential power to rule the world.

PART 3

WHO WEARS THIS PRIME NUMBER CROSS?

HER MAJESTY THE QUEEN

This portrait of Her Majesty The Queen was painted by Mr. Leonard Boden in 1968. Her Majesty is shown in the robes of the Sovereign Head of the Order of St. John and wearing the Insignia of the Order which were made for Queen Victoria. On November 28th, 1968, Her Majesty visited St. John's Gate, Clerkenwell, where the portrait was on view for the first time.

Fig 14

The Queen of England is adorned with the Templar Cross,

and thus has claimed this Sacred Geometry.

(Image taken from: "The Order Of St John" by E.D. Renwick, O.St.J. 1958)

H.M. The Queen, accompanied by H.R.H. The Grand Prior, at
The Royal Review, Hyde Park, 1956.

Fig 14a

The Grand Prior (walking beside the Queen), 1956,
proudly wearing the Templar symbol on his jacket.

I found a British One Pound note and looked closely at the Queen's crown. Here a close-up of the Queen's crown, and below the actual whole note.

Fig 14b
**Prime Number Cross visible
on Queen Elizabeth 11 of England's crown**

Being enchanted by such an observation, I decided to pen a drawing of this. In Fib 14c below, I have typed out the hand-written writing at the base of the drawing.

"The Queen of England Adorned with the Prime Number Cross" art by JAIN 20-8-08 Tanna Island. Copied from a One Pound note that has Sir Isaac Newton on back. Nb: these 2 crosses are on her Crown (Chakra). why? She rules the World as she understands the ancient mathematical mysteries. Nb: also there are 3 spirals like Sixes = 666 and below are 18 beads in her hair spaced every six beads = 6+6+6. She is thus Empowered.

Fig 14c
"The Queen of England adorned with the Prime Number Cross" on her crown. Art by Jain 20-8-2008, Tanna Island, Vanuatu.

Copied from a One Pound note that has Sir Isaac Newton on the reverse side.
nb: these 2 crosses are on her Crown Chakra! Why? She rules this Material World as she understands the ancient mathematical mysteries.
nb: also there are 3 spiral-like Sixes = 666 and below are 18 beads in her hair spaced every six beads = 6+6+6.
She is thus Empowered by the Harmonics of Creation.

Just for the record, you will see this symbol everywhere. I am including a few more photos of Kaiser Wilhelm 2nd of Germany.

Fig 15
Kaiser Wilhelm 11 of Germany
wearing his royal Prime Number Cross in full regalia.
The Maltese Cross is visible on his Heart chakra.

Germany (1867–1919).	Germany (1937–1945).
German Flag 1867 to 1919 (this image and the next, are taken from "Flags and Heraldry")	German flag with combined symbols of the Swaztika and Maltese Cross used in World War 2
German Imperial Flag	Flag for the Prussian Crown Prince

Fig 16
**Flags of Germany at different eras showing
the distinct symbolism of the Templar Cross**

PART 4

PHI'S PALINDROMIC PRIMES

PPP

— This connection between Prime Numbers and the Fibonacci Numbers / Divine Proportion, was first published in my previous book: The Book Of Phi, volume 4.

PHI CODE OF PRIMES (PALINDROMIC)

7 7 5 3 1 7 1 3 5 7 7 0

How is this Palindromic (in the Greek language, literally "running backwards" or reversed) Sequence derived?

We begin with the famous ancient Pattern of Phi Code 1: the Phi Code of 24 Repeating Pattern based on the digital compression of the fibonacci series:

1 1 2 3 5 8 4 3 7 1 8 9 8 8 7 6 4 1 5 6 2 8 1 9

and rearrange it into the familiar two rows of 12 digits:

1 1 2 3 5 8 4 3 7 1 8 9
8 8 7 6 4 1 5 6 2 8 1 9

And then examine the 12 Pairs of 9.

We then investigate the differences between the two rows, whether it be we subtract the Lower Row of 12 from the Higher Row of 12 or vice versa.
Either way, we end up with a Sequence of 12 digits that reads the same as it does backwards:

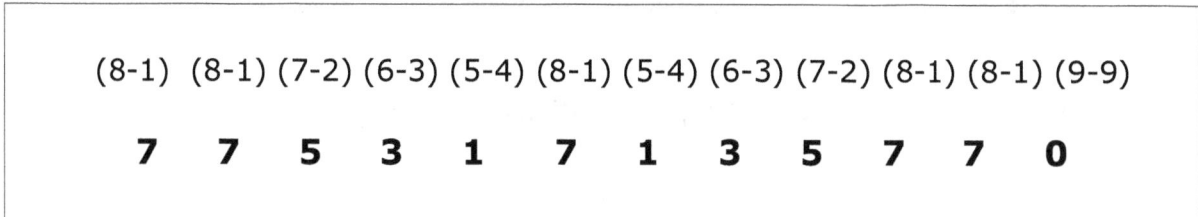

(8-1)	(8-1)	(7-2)	(6-3)	(5-4)	(8-1)	(5-4)	(6-3)	(7-2)	(8-1)	(8-1)	(9-9)
7	**7**	**5**	**3**	**1**	**7**	**1**	**3**	**5**	**7**	**7**	**0**

Fig 17
The Sequence of Prime Numbers hidden in the Phi Code 108.
It is now called "The Phi Prime Connection".

So here it is, a simple sequence of prime numbers based on Nature's Fibonacci Numbers, never before published material. Such a sequence of Prime Numbers could be translated into musical patterns.

CHAPTER 6

THE ATOMIC ART
OF THE MAGIC SQUARE OF 3X3

CHAPTER CONTENTS

PART 1

The TESSELLATED MAGIC SQUARE OF 3X3
CREATING The ATOMIC STRUCTURE
Of DIAMOND LATTICE

— How to Construct a Magic Square of 3x3 Yantram
— How to Tessellate 81 times or Tile the Magic Square of 3x3 with alternate tilts at 90°.
— Annie Besant and Charles Leadbeater, two clairvoyant Theosophists.
— Atomic Structure of Diamond Lattice, in 2-Dimensional Form

PART 2

SOLOMON'S SEAL Or STAR Of SOLOMON
DERIVED From The MAGIC SQUARE Of 3X3

— The Natural Square of 3x3 and Row By Row Analysis.
— The 3 Rows of the Magic Square plotted individually onto a Matrix of 9 Dots.
— The Pattern of 3 Rows super-imposed at right angles (90°).
— Charles Leadbeater's "Micro-Psi View" or image of Beryllium Crystal.
— The Seal of Solomon @ 0° + 90° Yantram.
— The Pentacle of Rabbi Solomon The King of Jerusalem.

PART 1

The TESSELLATED MAGIC SQUARE Of 3X3
CREATING The ATOMIC STRUCTURE
Of DIAMOND LATTICE

Here is an amazing correlation with Magic Squares and Atomic Art. We first need to define and learn how to construct a Magic Square of 3x3 Yantram.

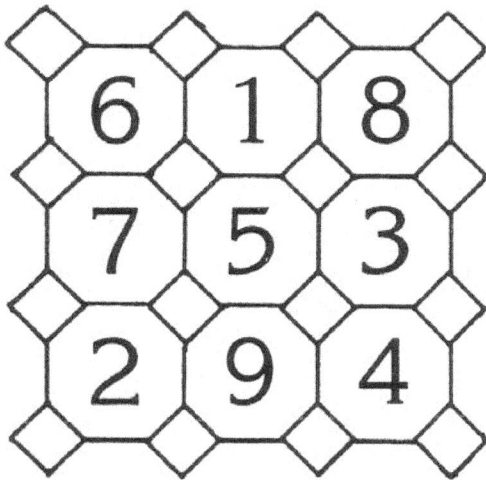

Fig 1a

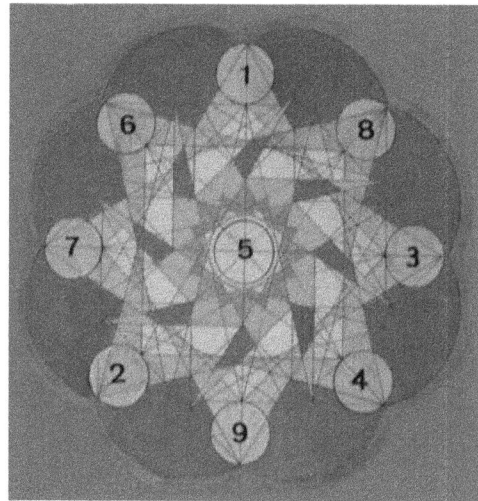

Fig 1b

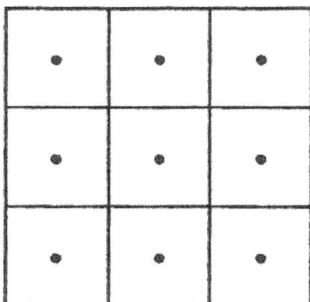 

Fig 1c

Fig 1

How to Construct a Magic Square of 3x3 Yantram.

The 9 numbers of the Lo-Shu are represented by the 9 dots.

Fig 1 shows one of 8 possible arrangements for the Lo-Shu. Fig 1a is just the numbers, and Fig 1b is a creative placement of these same numbers inserted into a Magic Square of 5 pattern, hand-drawn and created by Jain (from one of my early books: "Join The Dots").

Fig 1c shows that the 9 numbers are now represented as dots at the centre of the 9 cells.

To Translate Number into Art, Fig 2, symbolically looking for Order amidst the apparent Chaos, we draw a long unbroken line from 1 to 2 to 3 to the last number 9, and connect the circuit by joining a final line from 9 to 1, the Omega back to Alpha.

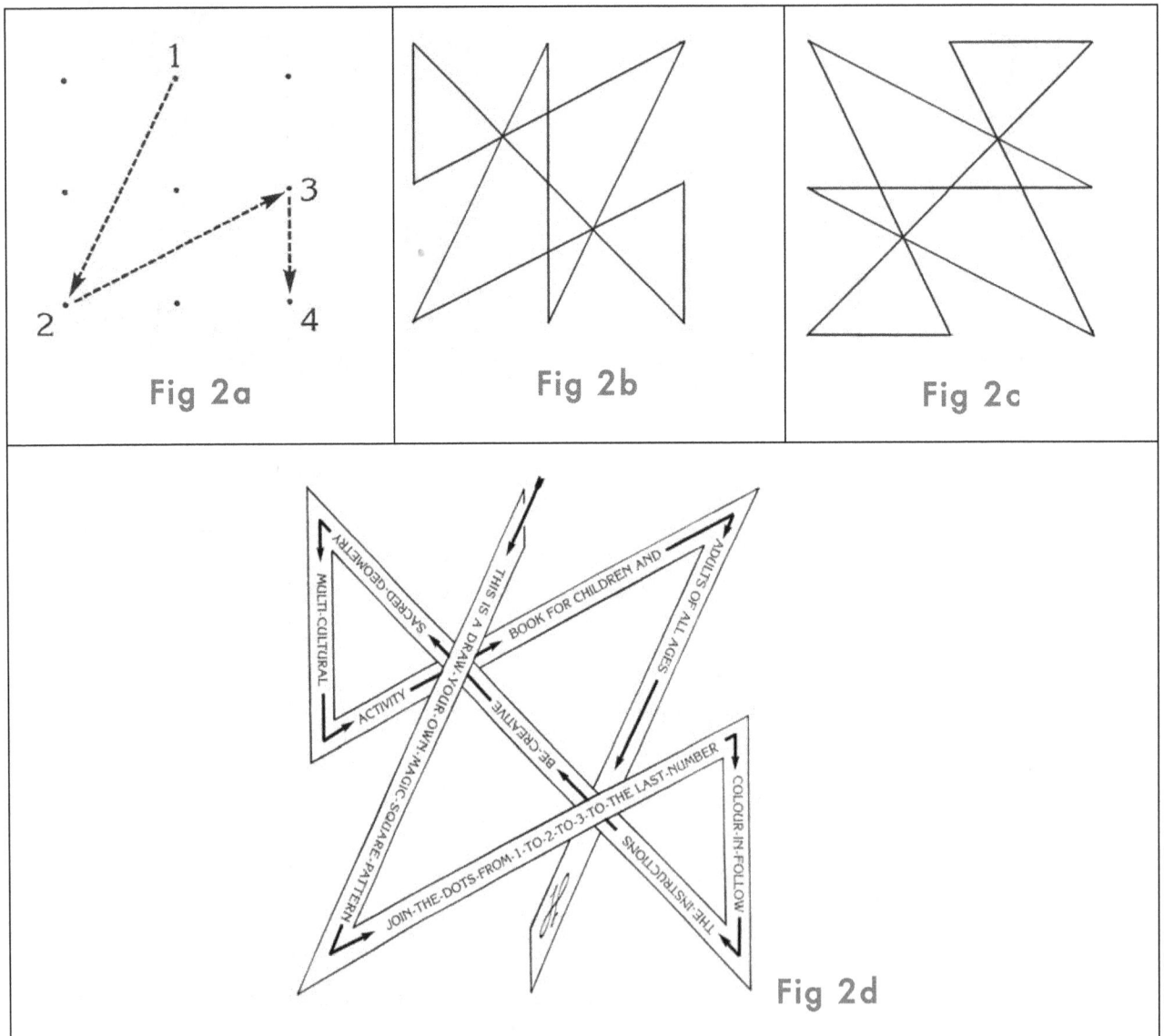

Fig 2a

Fig 2b

Fig 2c

Fig 2d

Fig 2

Creating the Magic Square of 3x3 Yantram

Fig 2a shows how we begin from the number 1 to 2 to 3 etc.

Fig 2b is the completed pattern with the final connecting line from 9 to 1. Technically this is called the Magic Square of 3x3 Yantram @ 0°, ie: as it is, in its pure, untransformed, non-rotated form.

Fig 2c shows the completed pattern but it has been rotated clockwise or anti-clockwise @ 9°, either way, it makes the same pattern.

Technically this is called the Magic Square of 3x3 Yantram rotated @ 90°.

Fig 2d is the uncompleted form of Fig 2b and has some fanciful writing, taken from "Join The Dots". It reads: "This Is A Draw Your Own Magic Square Pattern, Join The Dots From 1 To 2 To 3 To The Last Number, Colour-In, Follow The Instructions, Be Creative, Sacred Geometry, Multi-Cultural, Activity Book For Children And Adults Of All Ages".

As we prepare to Tessellate or Tile the Magic Square, we will be only interested in Figs 2b and 2c above. Fig 3 below shows how we alternate the tilt of the same magic square alternatively at 0° and 90°.

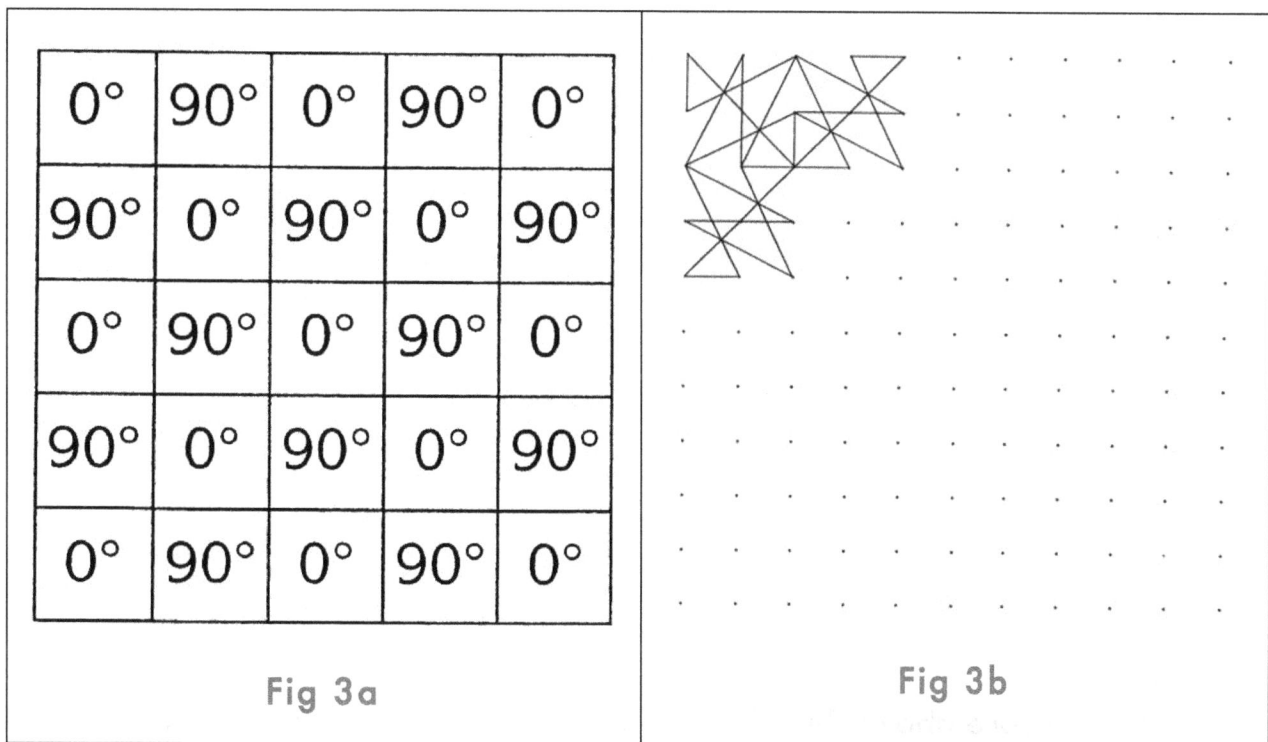

0°	90°	0°	90°	0°
90°	0°	90°	0°	90°
0°	90°	0°	90°	0°
90°	0°	90°	0°	90°
0°	90°	0°	90°	0°

Fig 3a

Fig 3b

Fig 3

Instructions how to rotate the Magic Square of 3x3 to be Tessellated or Tiled 25 times, with alternate tilts at 90°.

The first 3 patterns have been done to show how the pattern progresses.

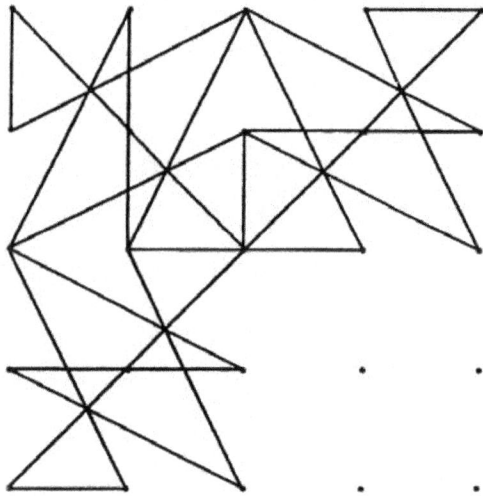

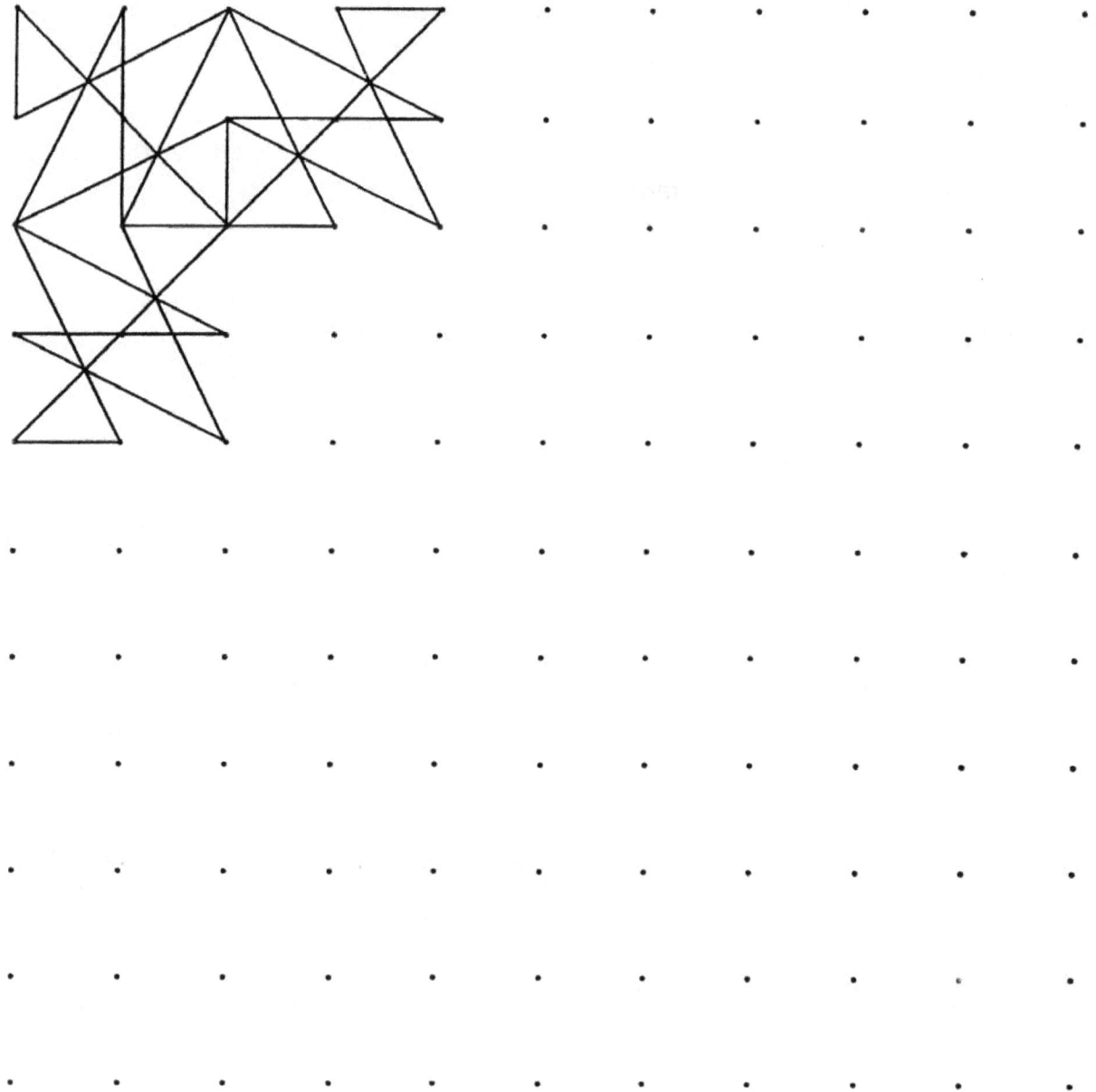

Fig 3c

Worksheet:

In the space above, Tessellate 25 times, the Magic Square of 3x3, using alternate tilts at 90°.

Obey the instructions given in Fib 3b, ie: draw the magic square at 0° then 90° then 0° then 90° etc.

The first 3 patterns have been done to show how the pattern progresses.

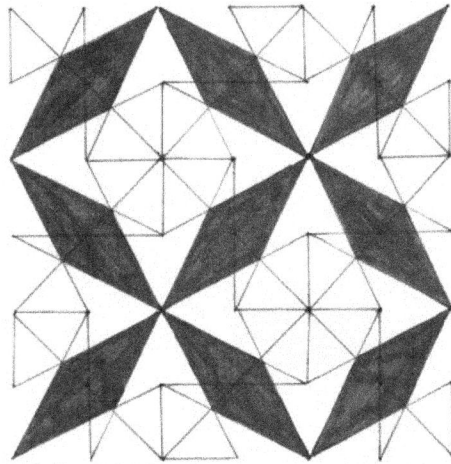

Fig 4a

The Magic Square of 3x3 being Tessellated or Tiled 9 times, with alternate tilts at 90°.

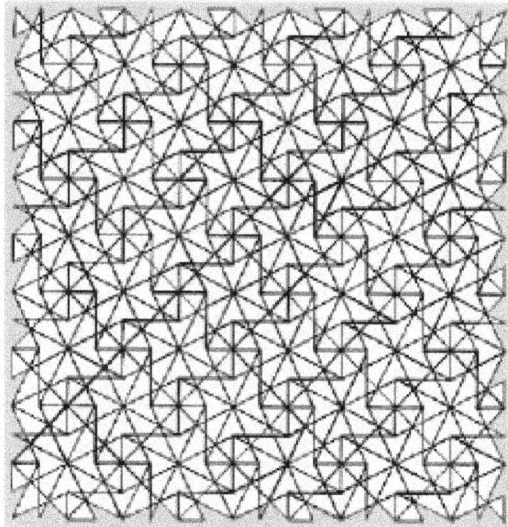

Fig 4b

The Magic Square of 3x3 being
Tessellated or Tiled 81 times, with
alternate tilts at 90°.

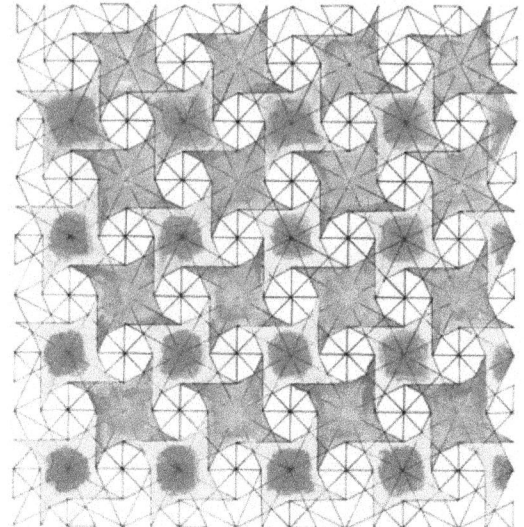

Fig 4c

Same pattern as Fig 4b but
coloured-in by creative isolation.

Fig 4

The Magic Square of 3x3 being Tessellated or Tiled 9 and 81 times,

with alternate tilts at 90°.

We are mainly interested in Fig 4b above, and we will make the stark connection with this pattern and the shadow form of the Atomic Crystalline Structure of Diamond Lattice.

Firstly, we need to introduce two remarkable Theosophists who lived in India over 100 years ago, Charles Leadbeater and Annie Besant. Both were able to go into the atomic world and actually draw what they saw! This is referred to as "Micro-Psi Art" and has been reproduced from "Occult Chemistry".

(First Edition was in 1908. Sub-Titled: Investigations by Clairvoyant Magnification into the Structure of the Atoms of the Periodic Table and of some Compounds).

Fig 5
Annie Besant and Charles Leadbeater, two clairvoyant Theosophists
(whose work currently clashes with the view by Quantum Physics
who smash particles at high speed; the cyclotron has 27 kilometers
circumference of tunnels underground in Switzerland,
to get a haphazard idea what an atom may look like).

The 2 diagrams below shows 2 remarkably similar patterns, the first one (Fig 4b) drawn by a child using magic squares artforms, and the second one (Fig 6), derived clairvoyantly of Diamond Lattice.

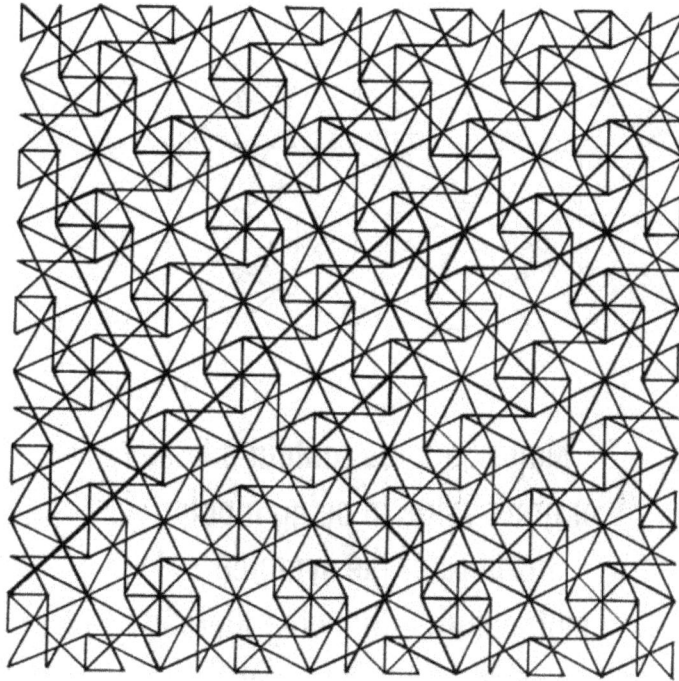

Fig 4b

The Tessellated Magic Square of 3x3
Tiled 81 times, with alternate tilts at 90°.

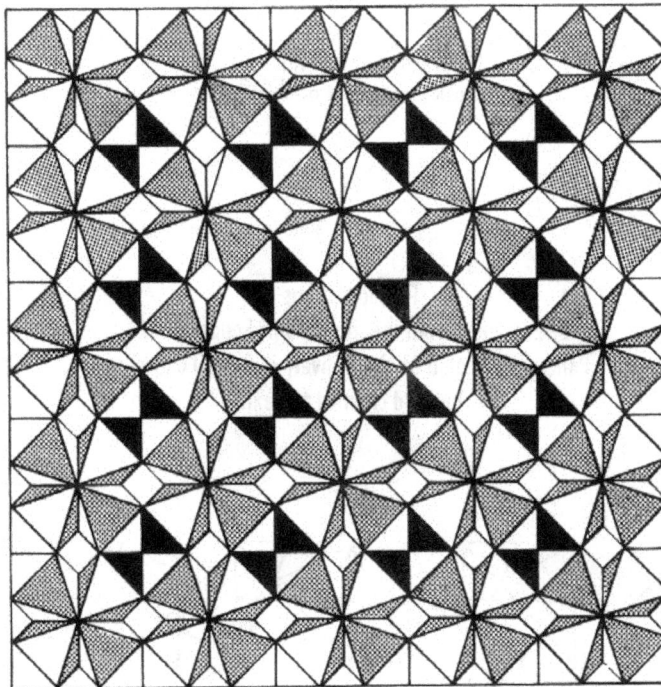

Fig 6

Atomic Structure of Diamond Lattice, in 2-Dimensional Form,
seen clairvoyantly by Charles Leadbeater and Annie Besant.

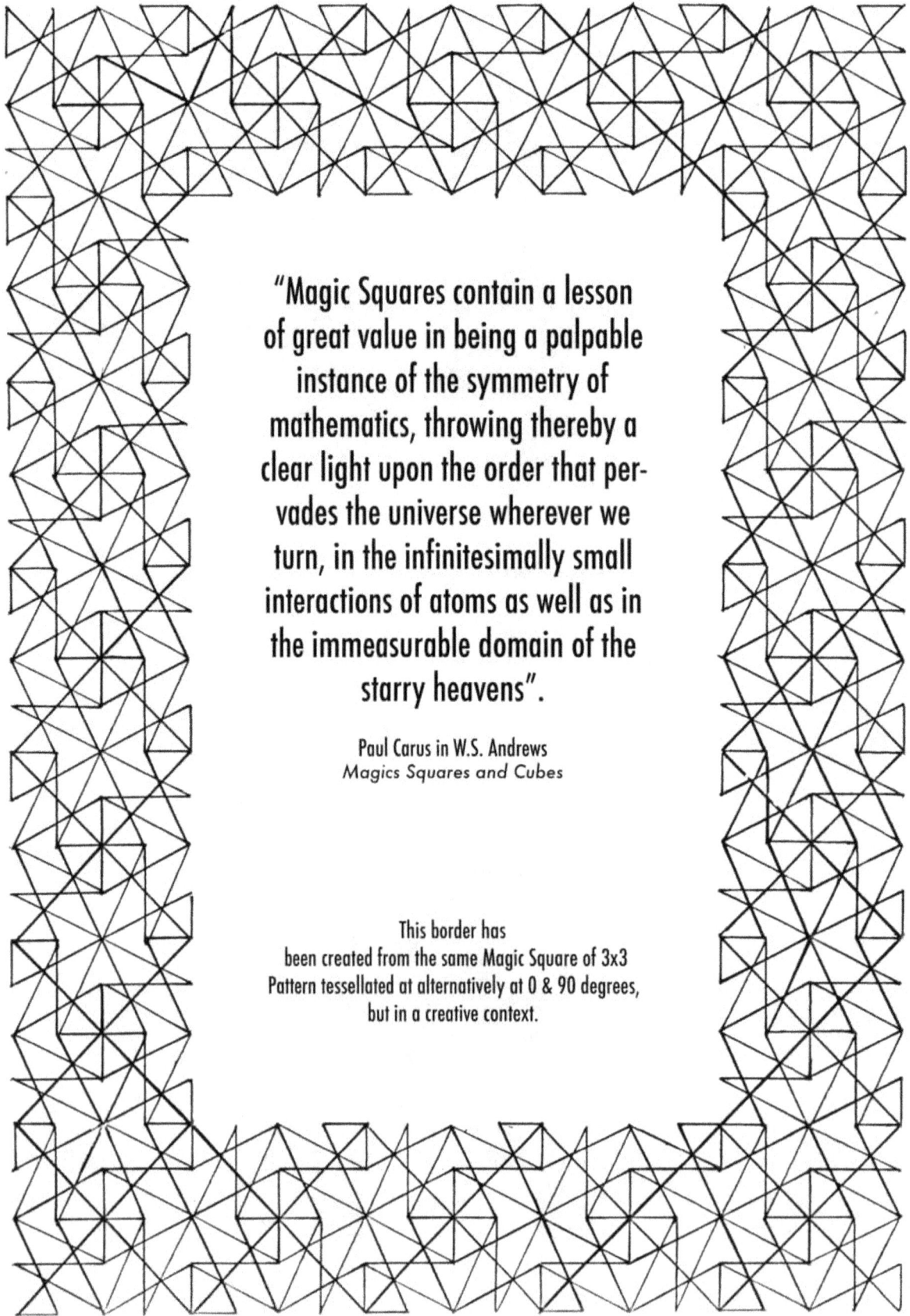

"Magic Squares contain a lesson of great value in being a palpable instance of the symmetry of mathematics, throwing thereby a clear light upon the order that pervades the universe wherever we turn, in the infinitesimally small interactions of atoms as well as in the immeasurable domain of the starry heavens".

Paul Carus in W.S. Andrews
Magics Squares and Cubes

This border has
been created from the same Magic Square of 3x3
Pattern tessellated at alternatively at 0 & 90 degrees,
but in a creative context.

This is to Certify that

....

has completed a
JAIN MATHEMAGICS FOR TEENS
CERTIFICATE COURSE
(in Vedic Mathematics)
During the period of
May and June of 2006
Comprised of 7 consecutive Saturdays
from 10am to 3pm at 777 Left Bank rd,
Mullumbimby Creek, 2482, NSW Australia

Congratulations on your highly improved abilities
in Mental, One-Line Arithmetic
and vastly increased Confidence and Memory Powers.

....

JAIN
Principal of the
JAIN MATHEMAGICS CURRICULUM
FOR THE GLOBAL SCHOOL

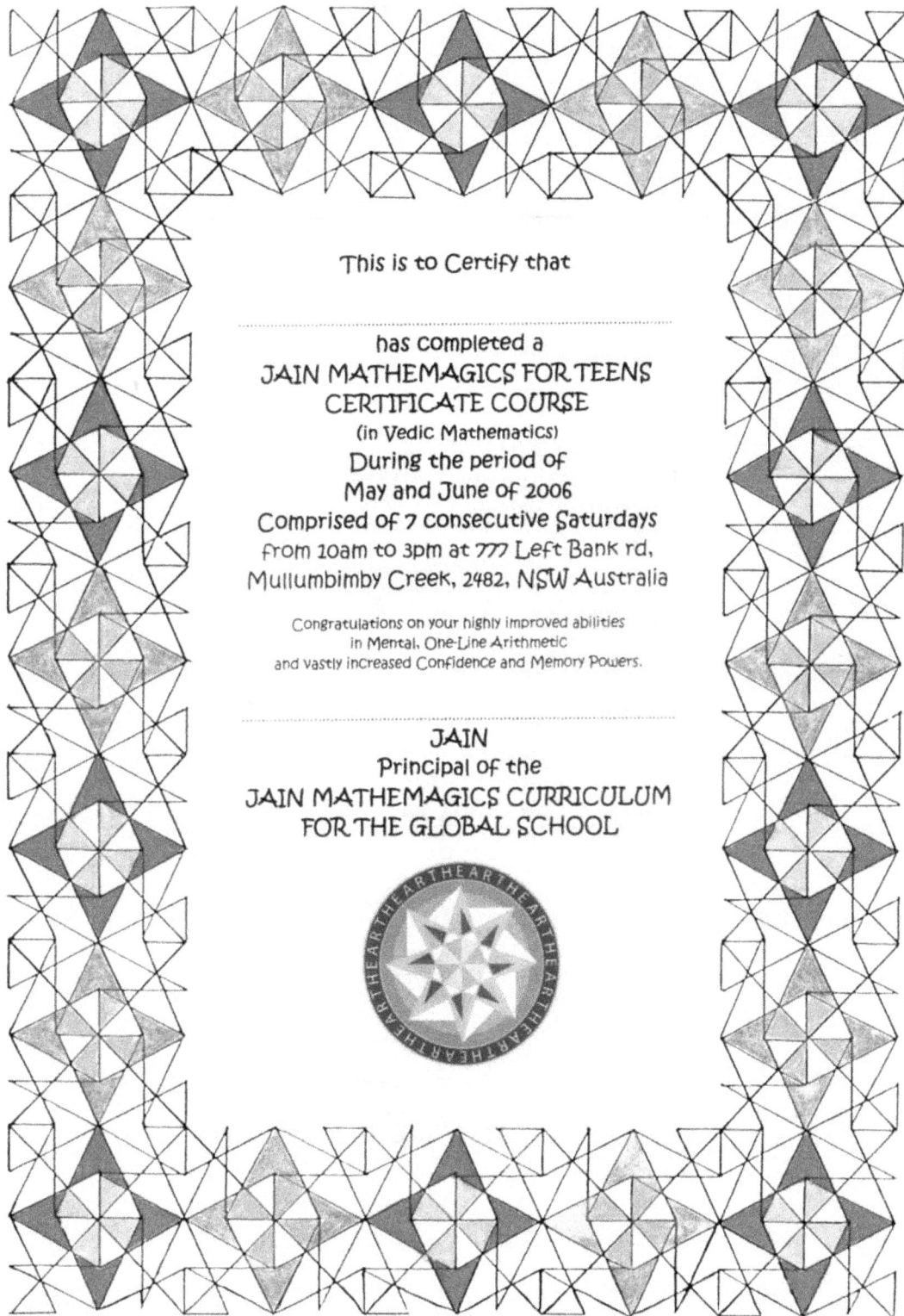

Fig 7b
The tessellated Magic Square of 3x3 used creatively as a border

PART 2

SOLOMON'S SEAL Or STAR Of SOLOMON DERIVED From The MAGIC SQUARE Of 3X3

ROW BY ROW ANALYSIS of the LO-SHU
(the Magic Square of 3x3)

Here is the traditional Magic Square of 3x3 having its 3 rows, 3 columns and 2 diagonals adding to 15:

Fig 1
The Magic Square of 3
Having all sums of columns, rows and diagonals to be 15.

(Footnote or Appendix: (Jain wearing a Tutenkamen mask and a Magic Square of 3x3 fluorescent robe created for a performance called: "**The Theatre of the Holy Numbers**" a concept of teaching Mathematics via Theatre and Art. Byron Bay, Epicentre, circa 1995 for the 2nd Global Mayan Conference where 700 people attended).

This Magic Square of 3 is only one of 8 possible arrangements. eg, The top row 6-1-8 could be at the bottom row, or down the side, or reading in reverse as 8-1-6 etc. These 8 permutations all produce the same Yantra or Pattern, when the Numbers are translated Into Art.

DID YOU KNOW?

6	1	8
7	5	3
2	9	4

2	9	4
7	5	3
6	1	8

6	7	2
1	5	9
8	3	4

2	7	6
9	5	1
4	3	8

These are the 8 possible Variations or Permutations of the Loh-Shu or the Magic Square of 3 x 3. They all have sums of 15.

8	1	6
3	5	7
4	9	2

4	9	2
3	5	7
8	1	6

8	3	4
1	5	9
6	7	2

4	3	8
9	5	1
2	7	6

Fig 2

The Magic Square of 3 has 8 possible permutations

Our goal is relate this magical arrangement of 9 numbers to its original state where the 9 numbers were in their natural counting order. This is called a Natural Square of 3x3 and appears in this form:

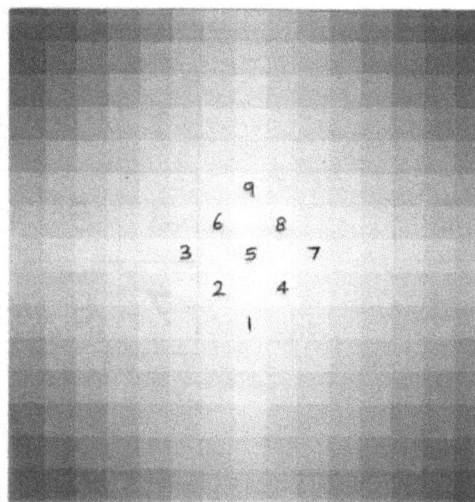

Fig 3

The Natural Square of 3 in its Diamond Form, showing the counting numbers from 1 to 9.

There is a special method how to create a Magic Square of 3x3 from the Diamond form of the Natural Square (called The Method of Transposition, where a central square is placed around the 4 corners of 6-8-2-4 and the 4 outer numbers 9-7-1-3 are moved 3 cells towards the centre or central cell of 5 and the Magic Square is mysteriously born).

The Natural Square is not "magic" because all the column, rows and diagonals do not add up to the same harmonic number.
The diamond form of the Natural Square can be rewritten into a plain square form:

1	2	3
4	5	6
7	8	9

Fig 3a
The Natural Square of 3
in its simplest square form.

Let us place the Magic Square of 3 side by side of the Natural Square of 3 so we can proceed to perform a certain operation.

6	1	8
7	5	3
2	9	4

Fig 1

1	2	3
4	5	6
7	8	9

Fig 3a

Fig 4
The Magic Square of 3 placed side by side
with The Natural Square of 3 for comparative purposes.

The question we ask is: "where do the numbers of the top row (6-1-8) of the Magic Square exist as positions in the Natural Square of 3?"

All these numbered squares can now be joined by lines to show their common connection.

To show this clearly, let 9 dots in a square form represent the 9 centres of the 9 numbers:

What would happen if we take each horizontal row of the MAGIC SQUARE OF 3x3 (fig 1), and identify where those triplets of numbers lie upon the NATURAL SQUARE OF 3x3 (fig 2). Solomon's Seal or Star of David is formed, another veritable form of ATOMIC ART (but with the added 7-5-3 axis identical to Beryllium's rotation about the centre of mass, in **Fig 6b**).

The Method to elucidate this phenomenon is by a process I call "ROW BY ROW ANALYSIS" of the 3 horizontal lines or rows of the Magic Square of 3x3 aka Lo-Shu.
Each of the 3 rows will be individually plotted into the Natural Square of Fig 2.

Row 1:   (6 — 1 — 8)
Row 2:   (7 — 5 — 3)
Row 3:   (2 — 9 — 4)

To turn Number Into Art, we need to allow each number or cell to be replaced symbolically by a "Dot" to form a simple 3x3 matrix of 9 Dots, these 9 dots are really the numbers of the Natural Square (seen several times in Fig 3).
Its like we are asking: What would the pattern look like for (6 - 1 - 8) upon the Natural Square (Fig 3a) etc.

When we examine the 3 rows we notice that two are triangular and 1 is merely a straight diagonal line, but it is this "straight diagonal line" that has metaphysical appeal, and will be shown in Fig 5. Also, notice that the two triangles of Figs 5a & 5c are elongated! (actually they are in the Root 5 ratio which is emblematic of the Phi Ratio / Golden Mean and this is shown in Fig 10).

6	1	8
7	5	3
2	9	4

Fig 1

The Magic Square of 3x3

1	2	3
4	5	6
7	8	9

Fig 3a

The Natural Square of 3x3

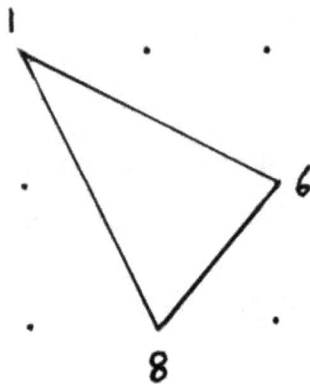

Fig 5a

Row 1: (6 – 1 – 8)

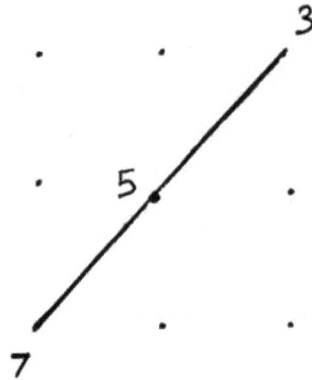

Fig 5b

Row 2: (7 – 5 – 3)

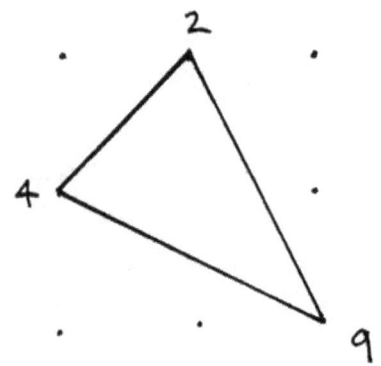

Fig 5c

Row 3: (2 – 9 – 4)

Fig 5

The 3 Rows of the Magic Square

plotted individually onto a Matrix of 9 Dots

"ROW BY ROW ANALYSIS" of the Magic Square of 3x3

Our next task is to ask: What would happen if we superimpose these 3 images upon themselves, as if they were drawn upon clear plastic transparencies. It will form the basic pattern in Fig 6.

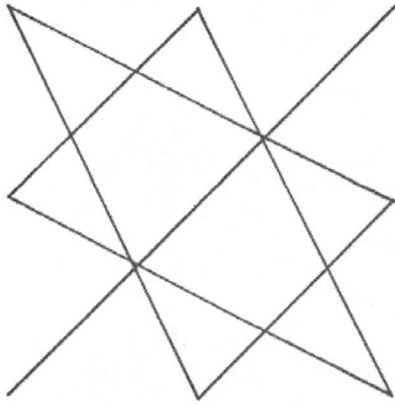

Fig 6a

The Super-Imposition of the 3 Rows

Forming an Elongated Star of David

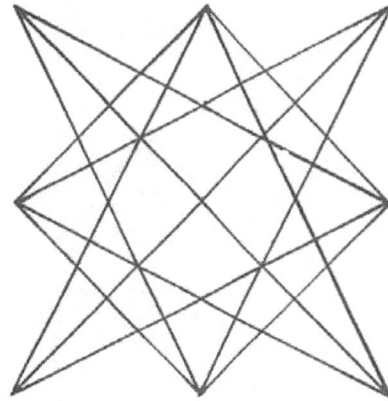

Fig 8a

The Pattern of 3 Rows super-imposed
at right angles (90°). See Fig 8.

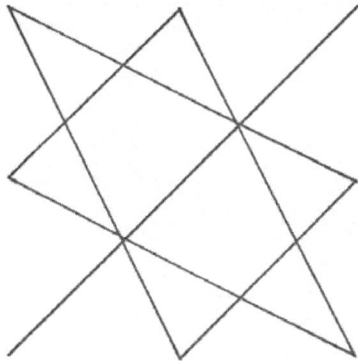

Fig 6a

Shadow form of Beryllium crystal

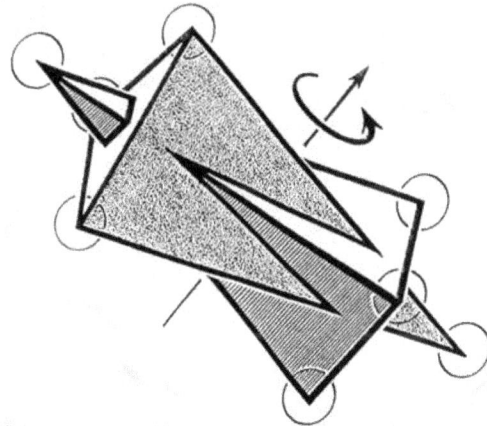

Fig 6b

3-Dimensional Form of Beryllium crystal
cropped from Fig 7

Fig 6

We have mathematically derived the crystalline shape of Beryllium (in 3-dim)
via Row By Row Analysis of the M. Sq. of 3x3. This is called "Transduction"
going from a higher 3-Dim form to its lower (2-dim) Shadow Form Fig 6a.

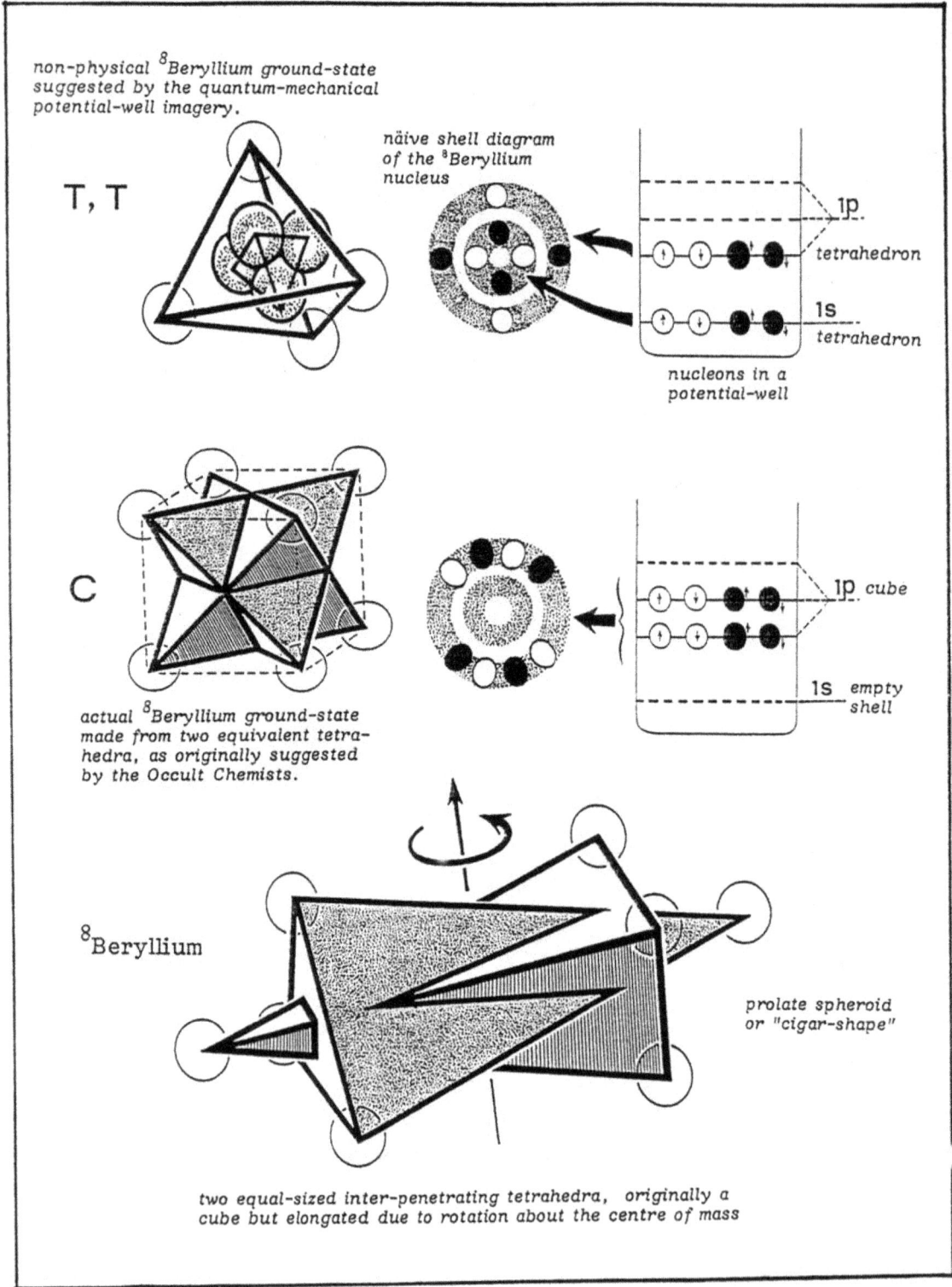

non-physical ^{8}Beryllium ground-state suggested by the quantum-mechanical potential-well imagery.

T, T

näive shell diagram of the ^{8}Beryllium nucleus

1p
tetrahedron

1s
tetrahedron

nucleons in a potential-well

C

actual ^{8}Beryllium ground-state made from two equivalent tetrahedra, as originally suggested by the Occult Chemists.

1p cube

1s empty shell

^{8}Beryllium

prolate spheroid or "cigar-shape"

two equal-sized inter-penetrating tetrahedra, originally a cube but elongated due to rotation about the centre of mass

Fig 7

This "Micro-Psi View" image is reprinted from Chris Illert's "Alchemy" who discoursed in full detail on Charles Leadbeater's out-of-print classic book: "Occult Chemistry" written over a hundred years ago.

This is my interpretation of the 3 main diagrams in Fib 7:
Leadbeater, with his rare "micro-psi" abilities, like remote-viewing, was able to describe in full detail the moons of Jupiter, without a telescope or modern technology over a hundred years ago. His book, called "Occult Chemistry" has been banned and out of print as its essential sacred geometry contradicts the current view of quantum physics.

This is a profound realization that a young student, merely by Joining The Dots, or utilizing The Art Of Number, can access the invisible world of Nuclear Geometry and Physics. This **Atomic Art** changes the consciousness of the viewer to comprehend that Mathematics is truly the basis of Creational Art in terms of it being the Universal Language of Shape and Pattern Recognition.

— The diagram on the top left is how Quantum Physics regards or views the atomic structure of the crystal Beryllium which is a small tetrahedron inside a larger tetrahedron. (Remember that Quantum Physics smash particles at high speeds in the Cyclotron which is destructive and dubious and the wrong way of thinking).
The top left diagram of Fig 7, where it says "T T" means a small tetrahedron inside a larger tetrahedron. That is the meaning of "T T".

— The diagram below this shows how Sacred Geometry or Occult Chemistry regards or views the atomic structure of the crystal Beryllium which has two inter-penetrating tetrahedra going through one another, overall forming a cube (Remember that Occult Chemists lucid dream or Soul travel to see the big picture non-destructively).
"C" in the diagram means "Cube". It is from clairvoyant observations, called "Micro-Psi" It shows the 3-dim Star of David known as Stella Octangula (Leonardo da Vinci's term for it) or Star Tetrahedron, both tetrahedrons being interlaced in a cubic array. Since the 8 vertices of this shape fit the corners of a cube, it is marked as: "C" for Cube. I subscribe to the latter, as did Kepler in the C16th who believed all atomic structure was based on the nesting of the 5 Platonic Solids.
Though we have two schools of thought, it is up to the reader to investigate these opinions and come up with their own conclusions based on their own research.

— The diagram at the bottom reads:
"Two equal-sized inter-penetrating tetrahedra, originally a cube but elongated due to rotation about the centre of mass".

I suspect these two Clairvoyant scientists, Charles Leadbeater and Madame Besant, were seeing two elongated tetrahedral in the phi ratio, that is if the equiangular triangular base sides are all 1 units, then the edge sides are 1.618...

Is Magic Square Art therefore a key to the backdoor of Nuclear Physics? Does my naïve diagram of the shadow of Beryllium being a shadow of the a 3-dimensional crystalline form indicate that Nuclear Physics, by smashing particles, are not seeing the real picture, that they have blundered.

Even the 2nd Row of numbers: (7 – 5 – 3) conforms exactly to the arrowed line in the base of Fig 6.

Can you see how this 3rd diagram is the shadow of this 3-Dimensional elongated Star Tetrahedron, (ie: it is exactly the same as The Seal of Solomon derived from the Magic Square of 3x3 which had its 3 rows analyzed and substituted in the Natural Square). In short, we have effectively created Atomic Art and Sacred Symbol, from the most simplest array of numbers: both the Natural Square and the Magic Square.

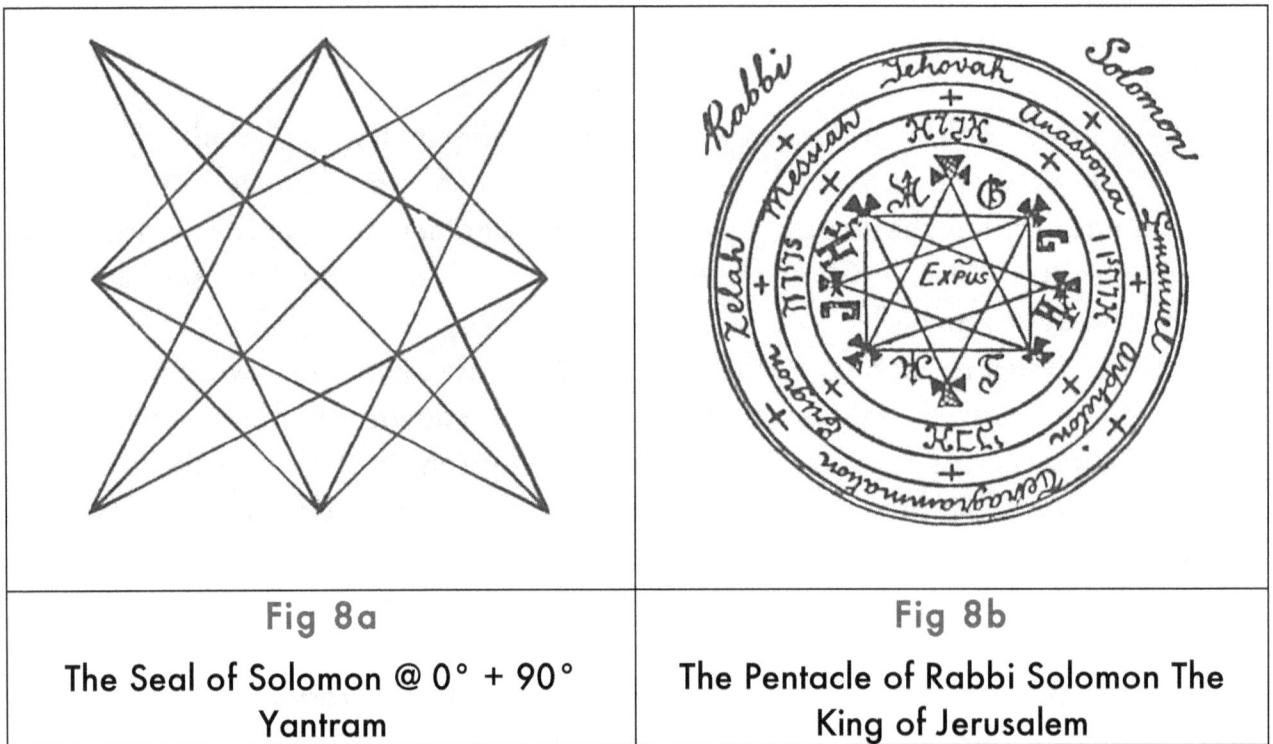

Fig 8a	Fig 8b
The Seal of Solomon @ 0° + 90° Yantram	The Pentacle of Rabbi Solomon The King of Jerusalem

Fig 8

The Seal Of Solomon (on right) is almost identical to the mathematically derived version from the Magic Square of 3x3 (on left).

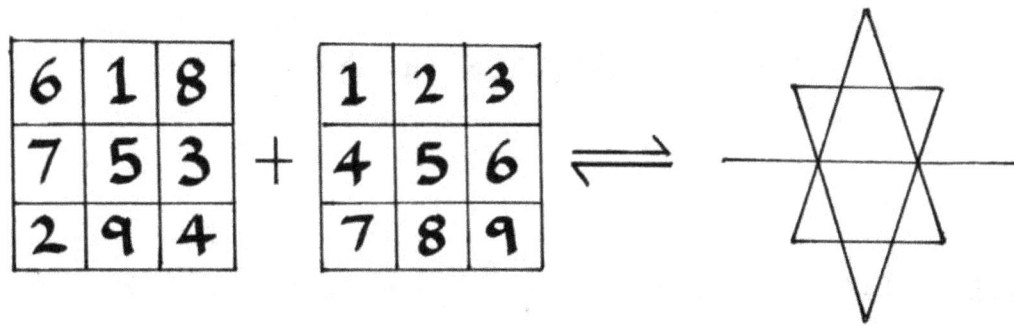

Fig 9

The 3 Row Patterns indexed against the Natural Square

This is a fanciful form or a summary of what we have hitherto accomplished:
We took the most simplest things in all the realms of mathematics: The Natural Square (you can't get any simpler than a square with numbers from 1 to 9 in sequential or consecutive natural counting order), plotted one against the other, and arrived at a sacred symbol.

When we pick up this image and super-impose it upon itself at 90°, it takes on a more symmetrical form appearing very closely to King Solomon's Seal.
(nb: the 2 main diagonals running through the centre have been removed.)

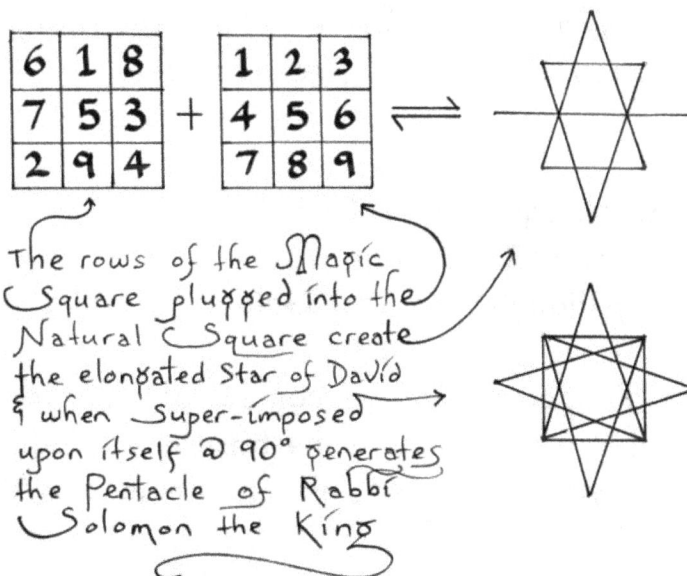

The rows of the Magic Square plugged into the Natural Square create the elongated Star of David & when Super-imposed upon itself @ 90° generates the Pentacle of Rabbi Solomon the King

Fig 9a

Summary of the Mathematical Derivation of Solomon's sigil.
Compare this now to Fig 8a, how similar these sigils appear.

189

Calculating long side of Elongated Triangle using Pythagoras' Theorem:

$$AC^2 = AB^2 + BC^2$$
$$= 1^2 + 2^2$$
$$= 5$$
$$AC = \sqrt{5} = 2.236\ldots$$

nb: $\sqrt{5}$ is part of the Phi formula $\dfrac{1+\sqrt{5}}{2} = \phi$

Fig 10

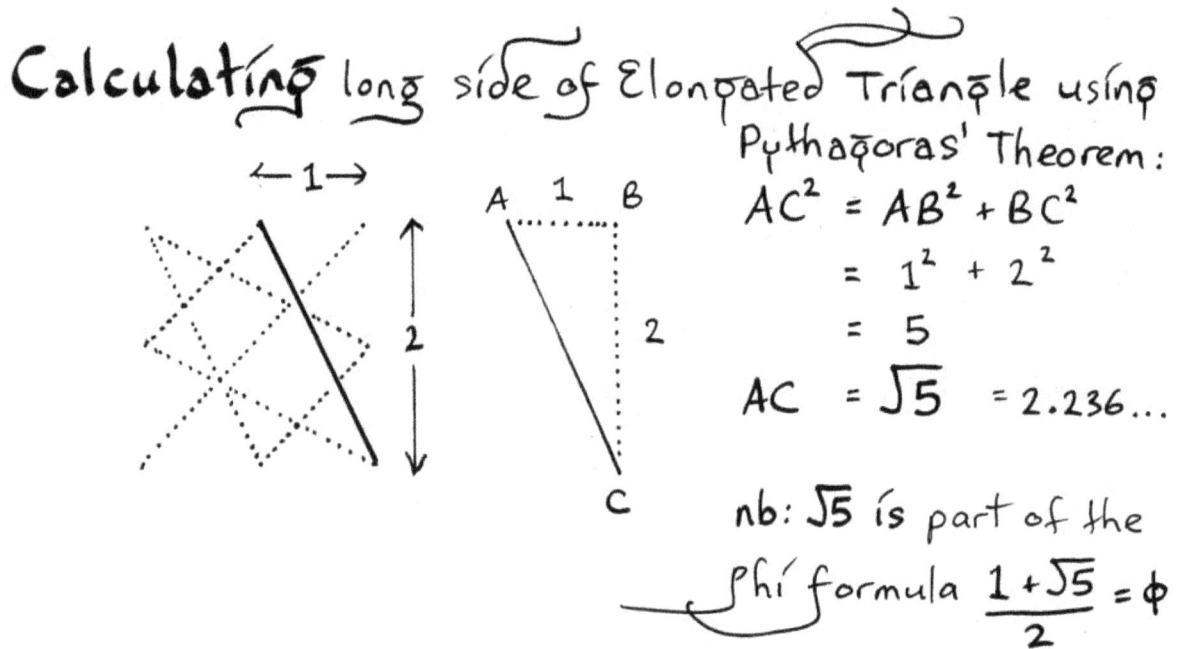

Root 5 Triangle in Solomon's Elongated Triangle
measured by Pythagoras' Theorem

Just for the record:

AB = 1 unit thus the whole square is really a 2x2 square

BC = Root 5 as calculated and shown above

The 2 interpenetrating triangles are not Golden Triangles (ie: not in the Phi Ratio).

The height of these elongated triangles can be determined from Pythagoras' Theorem, a right-angled triangle where the base is $\sqrt{2}/2$ (the Square Root 2 divided by 2) and the hypotenuse is $\sqrt{5}$, giving a height of $3/\sqrt{2} = 2.1213\ldots$

If the ratio of a Golden Phi Triangle is the Slope Length ÷ Base Length, let's test to see if the Seal of Solomon Elongated Triangles are in an anointed proportion.

$= \sqrt{5} \div \sqrt{2}$

$= 2.236\ldots \div 1.4142\ldots$

$= 1.5811\ldots$

This is very close to 1.618033... the Divine Proportion which is 1.62... when rounded off, but technically, we can conclude that these elongated triangles, are not Golden Triangles.

Thus the comparative relationship to Solomon's Elongated Triangle, and the Golden triangle is **1.58... : 1.62** which is so close to being the same, but it is a different frequency or vibration so we must isolate this fact and leave it at this stage.

So our enquiry provides a negative result, but this is part of our research to compile and index and record such data for future generations.
Although they are different numbers, there is a slight and interesting connection in the Root 5 length as it is an integral part in calculation the Phi Ratio whose formula incorporates this mystical dimension of Root 5.
The Phi Formula, embodied in the pentacle mathematics is: $(1+\sqrt{5}) \div 2$
(1 + **Root 5**) divided by 2.

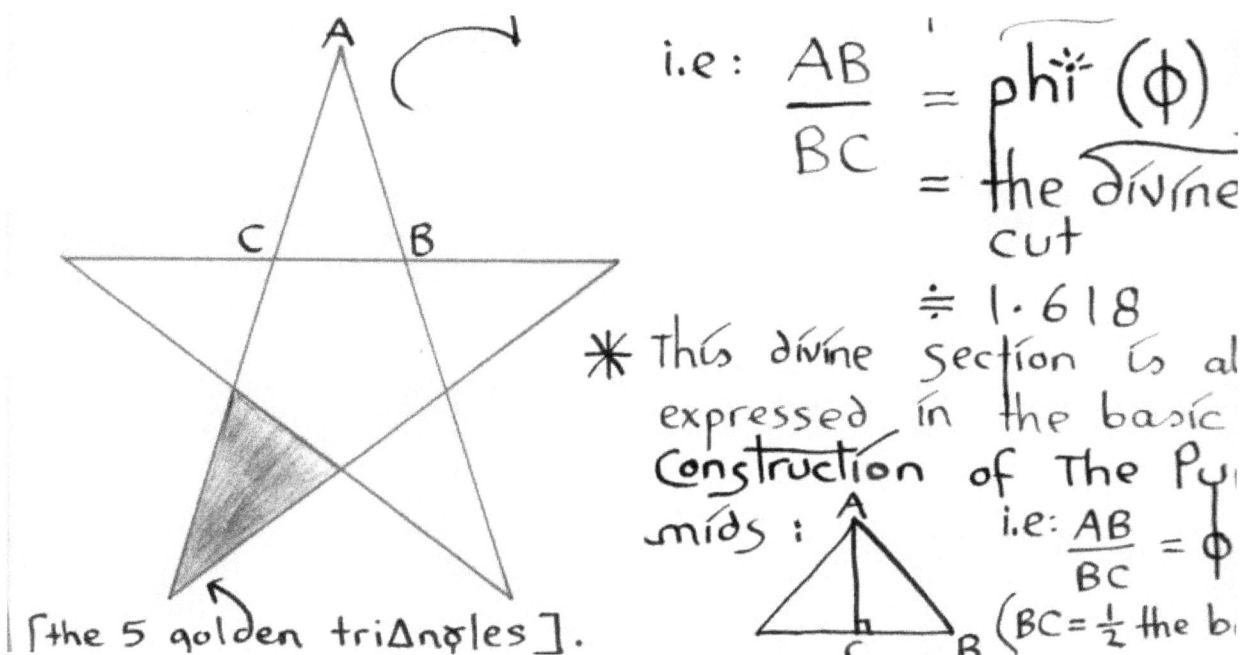

i.e : $\dfrac{AB}{BC}$ = phi (ϕ)
= the divine cut
$\doteq 1.618$

* This divine section is al expressed in the basic construction of The Py mids : i.e: $\dfrac{AB}{BC} = \phi$ ($BC = \frac{1}{2}$ the b

[the 5 golden triangles].

Fig 11
Depiction of the Golden Triangle 1:1.618033...
as one of the 5 arms of the Pentacle
(extracted from my earliest hand-written manuscripts
in the dimensions of the Tibetan style pages, drawn 1984).

The Elongated Triangle is seen again in the Magic Square of 3x3 in another variation when the 9 Numbers of the Magic Square are drawn in consecutive order ie: from 1 to 2 to 3 to the last number 9 giving this pattern:

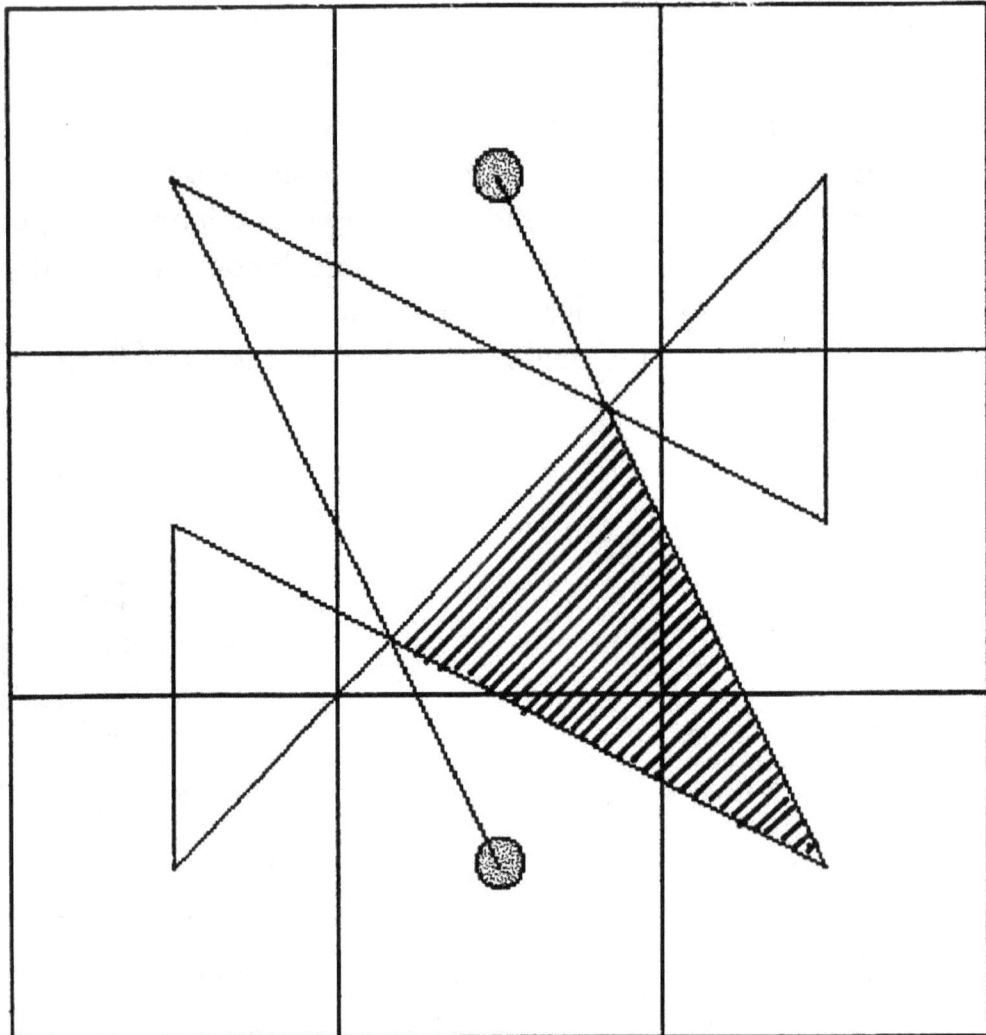

Fig 12
Depiction the Elongated Triangle as it appears
in the Magic Square of 3x3 Yantra at 0 degrees
This was shown in my early book:
THE BOOK OF MAGIC SQUARES volume 1 published in 1990

CONCLUSION:

Since the Golden Mean is the Mathematics of Aesthetical Beauty and Solomon's Seal approximates this ratio, as shown in the above diagram, then it becomes inviting and natural to appreciate the elegance of such a talisman.

What is of interest her is that the critical yantram Fig 9a that is identical to King Solomon's Seal or psychic emblem. A yantram, whose plural is "yantra" is a psychoactive device or power art or mystical diagram that conducts psychic essences.

Here we have a Pentacle of one of the greatest Rabbis or Kings in history, and his power glyph was based on the humble Magic Square of 3x3, which is the galactic mathematics of Base 9, creating the 6 pointed star and evolved from a 3 square.

This symbolism of the **3-6-9** was referred to by Tesla, the man who gave the world the wonders of Television, radar, radio, free energy systems etc is quoted as saying:

"IF YOU ONLY KNEW THE MAGNIFICENCE OF

THE THREE, THE SIX AND THE NINE...**3-6-9**...

THEN YOU WOULD HAVE A KEY TO THE UNIVERSE."

- NIKOLA TESLA

What follows are the original hand-written and hand-drawn pages that contained this foregoing information on the Star of Solomon based on the Magic Square of 3x3.

It comes from two sources:

SOURCE 1:
The next 4 images are sourced from The BOOK OF MAGIC SQUARES, volume 3, self-published 2000.

from THE GOSPEL OF THE
HOLY TWELVE:

The ⬤ST⭐R of SOLOMON.

can be arrived at from two different
Majik Square Sources:

Source 1: All Columns,
rows & diag-
onals have a Sum of 12
(that is why this M.Sq. is
included in this Chapter 12).
All opposing pairs are
double the Centre No.
The Squares shown in
Fig 129 are a 180° reflection of each
other. Their digits "3" "4" "5"
relate to the 'Pythagorean Right-
Angled Theorem $3^2 + 4^2 = 5^2$ as
shown in Fig 129.
Let us study the M.Sq. on the
right hand side of Fig 129.
In the space of Fig 130 below
connect the 3 points of "3" (a triangle).
Superimpose the 3 points of "4"
& then the
3 points of
"5" (a tri-
angle).
Then, upon
these 9
dots, do it
at 0° + 90°.

3	5	4
5	4	3
4	3	5

Cont.
Next
Page.

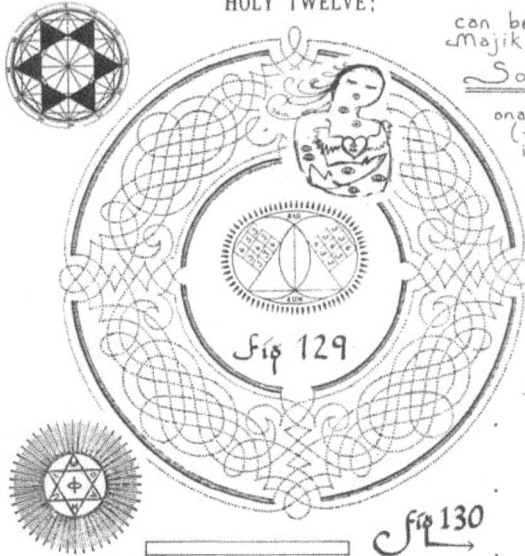

Fig 129

Fig 130

Page 1 of 4:

This page has extra information on a similar 3x3 square based on the numbers "3, 4 and 5". It happens to make the identical pattern as Fig 6a. This rare magic square based on 3,4,5 was sourced from "The Gospel of the Holy Twelve" a rare bible in my library (and available from my internet site).

Page 2 of 4:

CONT... "The ST⭐R of
SOLOMON Y⭐NTR⭐M"

OR "The Pentacle of King Rabbi Solomon".

Source 2:

Fig 131 →

(a) (b) (c)

6	1	8
7	5	3
2	9	4

Lo - Shu
or m.Sq.of 3

The Natural
Square of 3

1	2	3
4	5	6
7	8	9

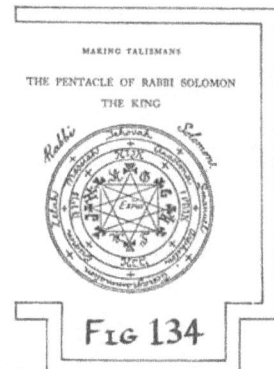

MAKING TALISMANS

THE PENTACLE OF RABBI SOLOMON
THE KING

FIG 134

Observe the 3 squares on your left
(Figs 131 a,b,c). Look at the top line of the M.Sq. of 3 which reads:
"6 1 8" then find where these 3 numbers are positioned in
the Natural Square of 3. The position of "618" in the
Natural Square is shown as Fig 131a
Fig 131b shows the row of "7 5 3" &
Fig 131c shows the row of "2 9 4".
In the space of Fig 132 superimpose the
← 3 diagrams of Fig 131 a+b+c to form an
elongated Star of Solomon (n.b: both triangles
are approximately in "Golden phi 1:1·618 Proportions). Then in the space of Fig 133 →
draw this Rabbi's Golden Star at 0°+90°. NB:
This Yantram is very similar to the ancient
diagram show above in Fig 134.

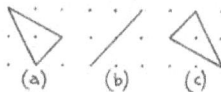
Page 149

FIG 132

FIG 133

Page 226

Even in the darkest period of mathematical learning, some people kept the science alive. Gerbert of Aquitaine (955–1003), who became the first French pope, Sylvester II, was a mathematician. He collected the surviving books of Latin and Greek authors to preserve their ideas. To help others do long and difficult calculations, he built globes of the earth and the heavens and popularised the use of the abacus.

FIG 30

SILVESTER II PAPA AQVITANVS

Fig 132

This Seal of Solomon is close to the Golden Mean Triangular shape. When tessellated the array is similar to the SILICON crystal field well known now as Atomic Art/ Structure!

(NB: The Solution for Fig 130 is identical to both Figs 132 & 133).

Fig 133

THE SACRED TREE OF THE SEPHIROTH

"A Great Truth is a Truth whose Opposite is also a Great Truth". Niels Bohr (Atomic Scientist).

ABSOLUTE KEY TO OCCULT SCIENCE

"We must continually count on the appearance of new facts, the inclusion of which within the compass of our earlier experience may require a revision of our fundamental concepts"
Niels Bohr

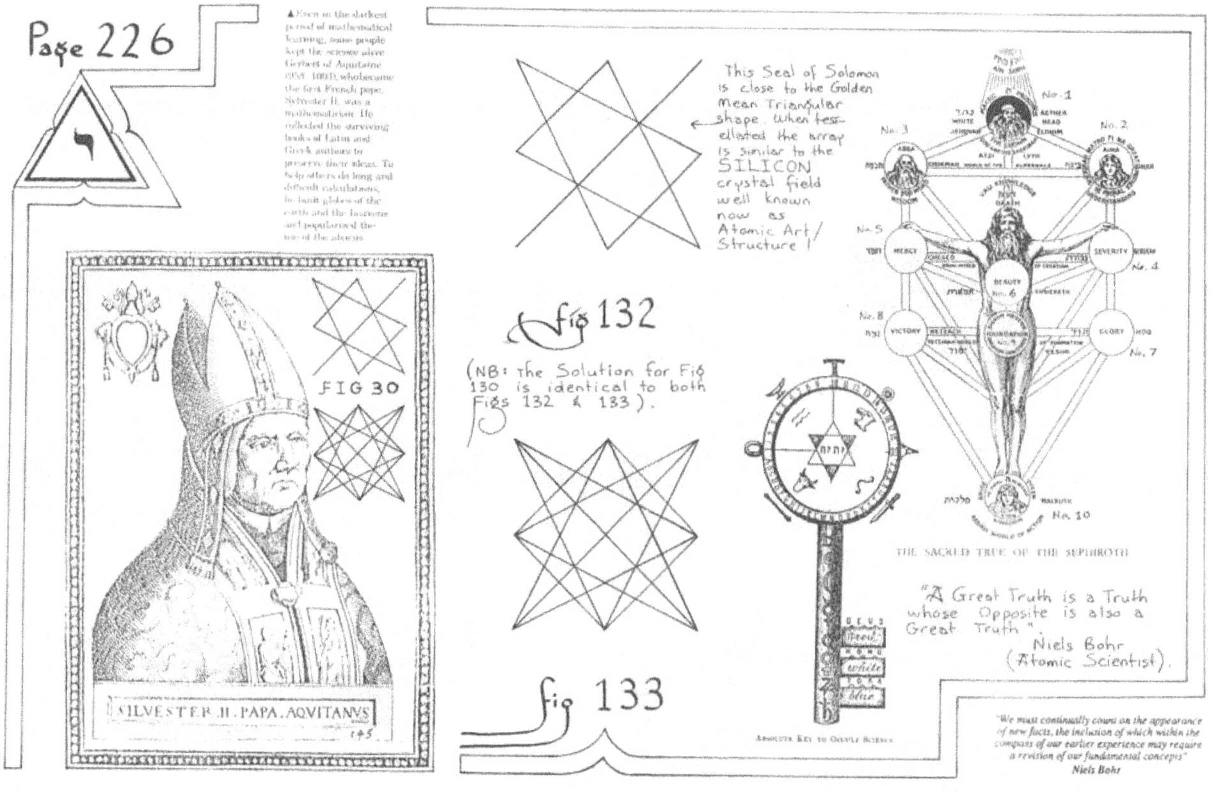

This is a Draw-Your-Own-
Majik-Square-Mandala-Color-In-
Activity-Book for Children + Adults.

Page 4 of 4:
The Star of Solomon
collaged by Jain above Krishna in one of his many divine forms:

196

SOURCE 2:

The next 3 images are sourced from THE BOOK OF PHI, volume 2, aka IN THE NEXT DIMENSION, self published 2003.

Page 1 of 3:

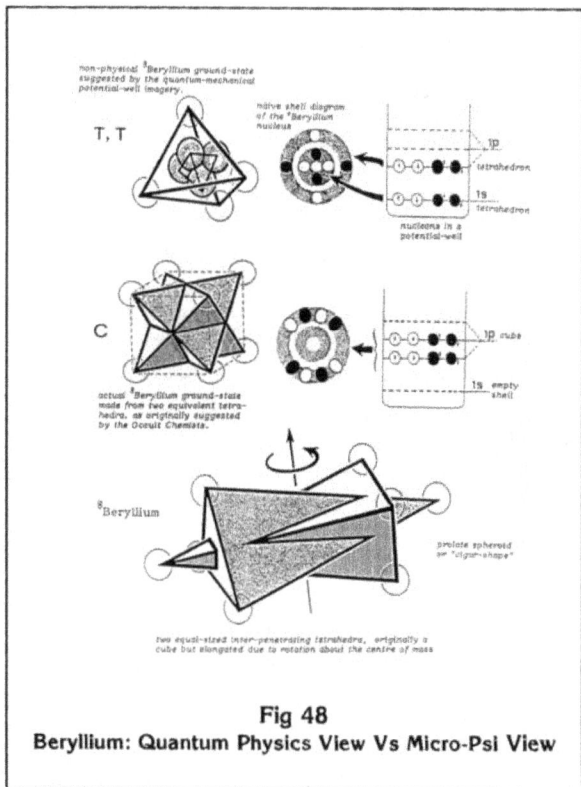

Fig 48
Beryllium: Quantum Physics View Vs Micro-Psi View

Figs 49a and 49b
Magic Square and Natural Square of 3 x 3

We are going to perform the task of "Row by Row Analysis" of each of the 3 horizontal lines or rows of the Magic Square of 3 x 3. Each row,
Row 1: [6 1 8], Row 2: [7 5 3] and Row 3: [2 9 4] will be plotted in the Natural Square. First let us represent either the Magic Square (M.Sq.) or Natural Square (Nat.Sq.) as a configuration of 9 cells or 9 dots in a 3 x 3 matrix. Figs 50 a,b,c shows the result of asking:
a)- Where do the 3 numbers of Row 1: [6 1 8] appear or position themselves in the Natural Square of 3 x 3?
b)- Where do the 3 numbers of Row 2: [7 5 3]

197

appear or position themselves in the Natural Square of 3 x 3?

c)- Where do the 3 numbers of Row 3: [2 9 4] appear or position themselves in the Natural Square of 3 x 3?

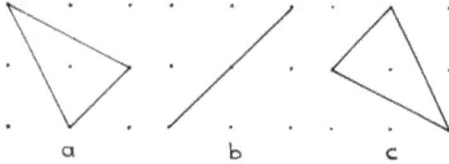

Figs 50 a, b, c
The Row-By-Row-Analysis of the M.Sq. of 3 Substituted Into the Nat.Sq. of 3.

So far we have 2 elongated triangles and a slanted diagonal line. What happens if we superimpose these three images upon themselves, as if they were drawn upon clear plastic transparencies.

Fig 51 shows the result of an elongated Star of Solomon that approximates the Golden Phi Ratio.

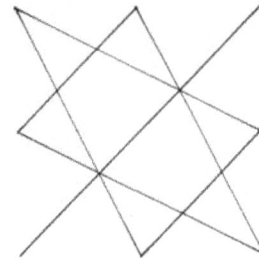

Fig 51
Star of Solomon Formed from M.Sq. of 3 x 3

Now if you were to look at Fig 51 with 3-D eyes you would perceive the two elongated phi triangles as two interpenetrating phi Tetrahedrons or phi Star-Tetra as shown in Fig 48, but also look at the slanted line in Fig 51 and correlate it exactly to Fig 48's spin of the cubic crystal Beryllium's elongation due to "rotation about the centre of mass". The similarity here between Magic Square Art and Nuclear Geometry can be no mere chance. If you drew Fig 51 again but superimposed the pattern upon itself at an angle of 90° you would be close to viewing King

Page 2 of 3:

Page 3 of 3:

Solomon's psychic Seal as shown in Fig 52a. Technically, in Magic Square Jargon, it is known as The Rabbi's Golden Star Yantram at 0° + 90°. ('Yantra' is the plural of 'Yantram', a Sanskrit word for "power diagram acting as 'transistor circuits' or channels conducting psychic essences"). Fig 52b shows "The Pentacle of Rabbi Solomon The King taken from an ancient bible of magical talismans.

Figs 52 a and b
M.Sq. 3x3 at 0° + 90°
Pentacle Of King Solomon

Fig 53: The ASHMOLEAN STONES

THE ASHMOLEAN STONES

Considered to be pre-Plato or some texts say 10,000 years old, these stones, currently in the Museum in Oxford, are hand-sized carvings of the 5 Platonic Solids and some Archimedean Solids, like the CubOctahedron. In Plato's "Timaeus": a cosmic dialogue based on these Platonic Solids, Tetra is Fire, Cube is Earth, Octa is Air, Icosa is Water, but of high importance is the FIFTH ELEMENT (Dodeca), its 12 faces being representative of Prana or the Universal Life Force or the **Phi Code of 12 Pairs**. Keith Critchlow (Magic Squareologist and Phi Architect and Lecturer) concludes that:
"The essential forms and numbers then act as the interface between the higher and lower realms".

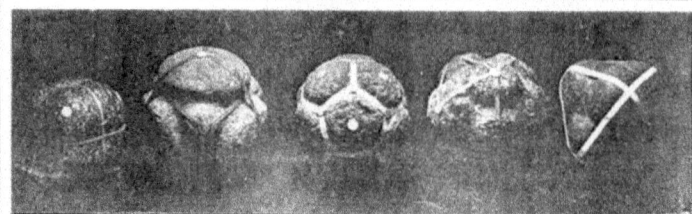

CHAPTER 7

THE STAR OF DAVID
DERIVED FROM THE NUMBERS 1 TO 7

The STAR of DAVID derived from the NUMBERS 1 to 7

THE 1 SEVEN

2   3

4

5   6

DIVINE 7 SPIRITS

CHAPTER SUMMARY:

- Ancient Puzzle
- Properties of the Star Of David Derived from the Numbers 1 to 7
- The fractal, ever recursive or repeating division of the triangle, known as the Koch Snowflake
- How to Convert Number Into Art
- The Coptic Cross
- The UniCursal Hexagram

Fig 1
The "Six Around The One".
The 6 centres of the 6 coins form the Star of David.

Here is an Ancient Puzzle.

Given 7 consecutive numbers (ie: one after the other in natural counting order), like the playing cards from 1 to 7, how can you arrange them onto a board where 6 of the numbers lie on a hexagon (circle divided into 6 parts and having 6 sides) and one number is in the centre, such that the 3 obvious columns have sums of 12 and the 3 opposing pairs have a sum which is double that of the centre?

You would need to pencil in the numbers into such a diagram:

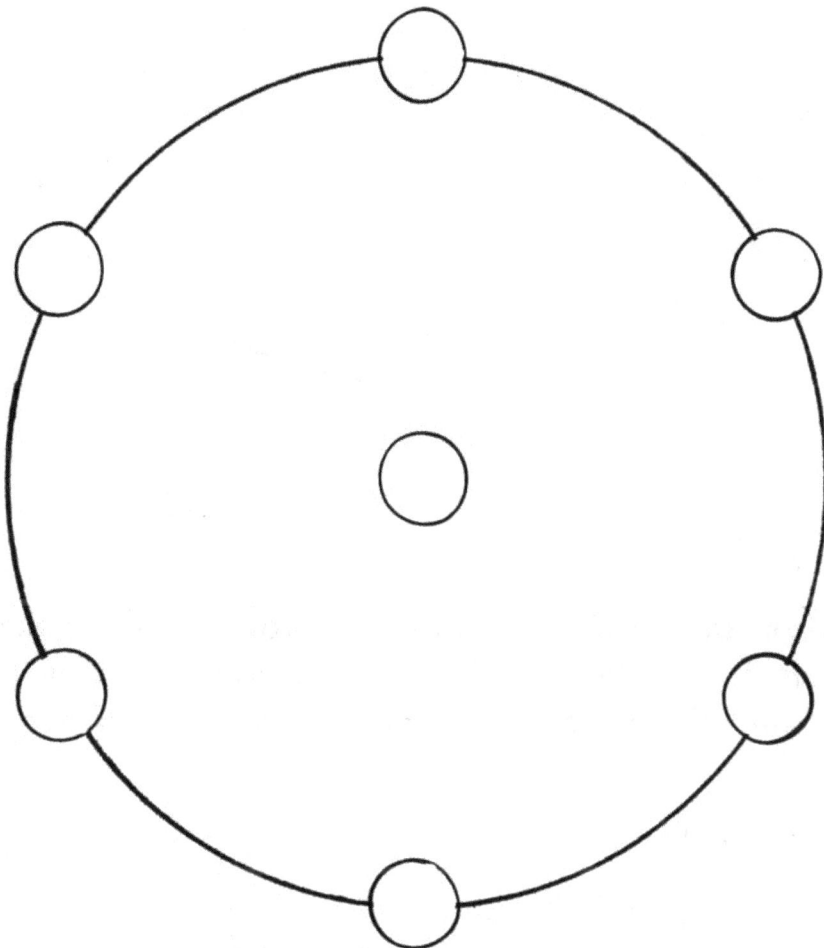

Fig 1a
Circular Board with Hexagonal Division
to write in the 7 consecutive numbers from 1 to 7

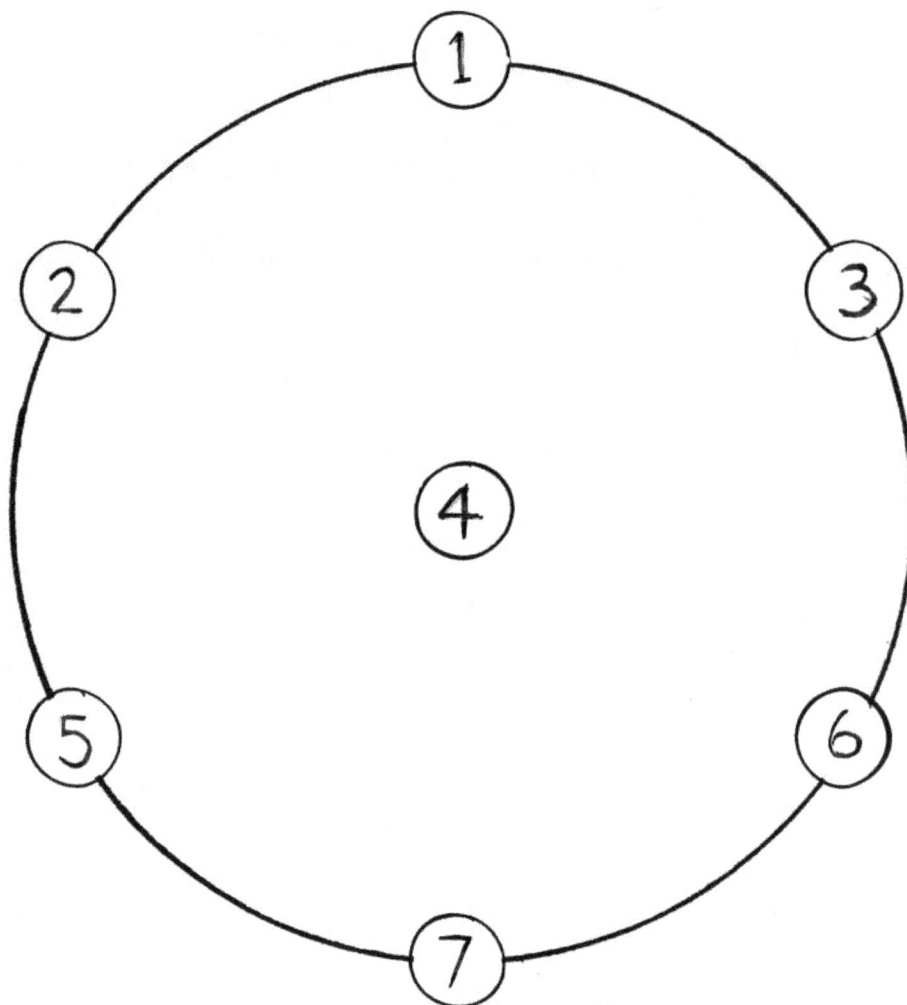

Fig 2
Solution for the placement of the consecutive numbers
from 1 to 7, where the columns have sums of 12
and the 3 opposing pairs have sums of 8.

Visualize the Path from 1 to 7 in terms Fluid Dynamics. Visualize the Circle as a Sphere. How would you connect the numbers from 1 to 7, in their natural counting order, using only curves? You can use the space in Fig 1 to practice this.
(Photocopy Fig 1 several times for this exercise).

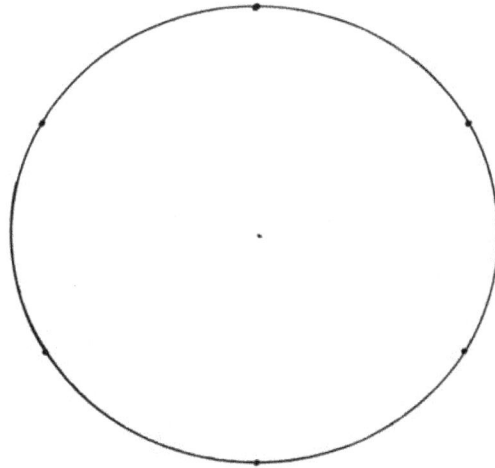

Fig 2a

**Grid for Using only curved lines
to connect the consecutive numbers
from 1 to 7 as if to simulate Fluid Dynamics**

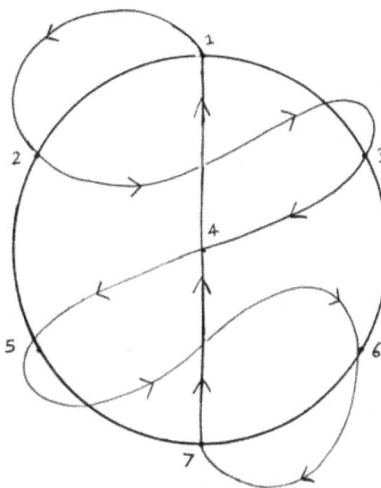

Fig 2b

**Using only curved lines to connect the consecutive numbers
from 1 to 7 as if to simulate Fluid Dynamics**

The Principle of Counter-Rotating Fields is inherently demonstrated in this elegant solution. As we draw the lines from 1 to 2 to 3 etc to the last number 7, symbolically looking for "Order Amidst The Chaos", notice how the curve or partial spiral reverses its spin, as it passes through the Centre. The central cell "4" is like a switchpoint or a cross-over point. In 3-Dimension it acts as the vacuum centre of the torus:

203

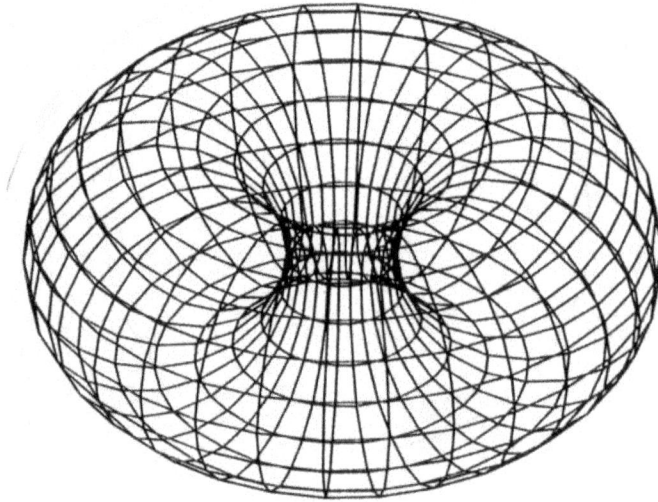

Fig 2c
**Torus has both clockwise and counter-clockwise fields,
an In and an Out in its Unified Field (Uni-Phi).
Thus the Torus knows how to unite duality.**

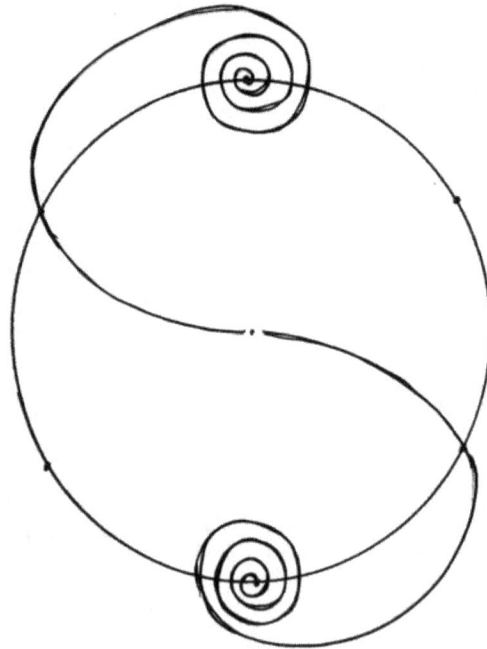

Fig 2d
The Double Torus is implied in this Star of David pattern
As the anti-clockwise spiral unwinds and passes through the centre, it reverses
it's spin and starts to go the other way, clockwise! It could suggest that the true
starting point for this yantram, is not "1" but "4" which lies in the centre.

PROPERTIES
of the STAR of DAVID
DERIVED from the NUMBERS 1 to 7

● The 3 Columns of 3 numbers have a Constant or Magic Sum of 12.
They include:

$1 + 4 + 7 = 12$

$2 + 4 + 6 = 12$

$3 + 4 + 5 = 12$

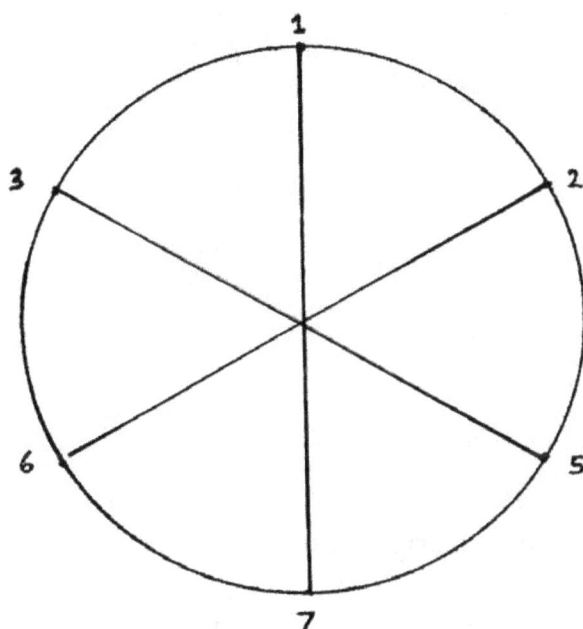

Fig 3a
Circular Diagram showing 3 axes
representing the 3 columns.
It is understood that the No. 4 is always in the middle.

● The 3 Opposing Pairs each add up to 8. They include:

$1 + 7$

$2 + 6$

$3 + 5$

Refer to Fig 3a to view these Opposing Pairs summing to 8.

●   These Pairs are double the central cell "4". No other number can occupy this central position, to make the pattern magic. Thus "4" is the pivotal point.

●   The 2 large main triangles have sums of 12. They include:
1 + 5 + 6 = **12**
2 + 3 + 7 = **12**

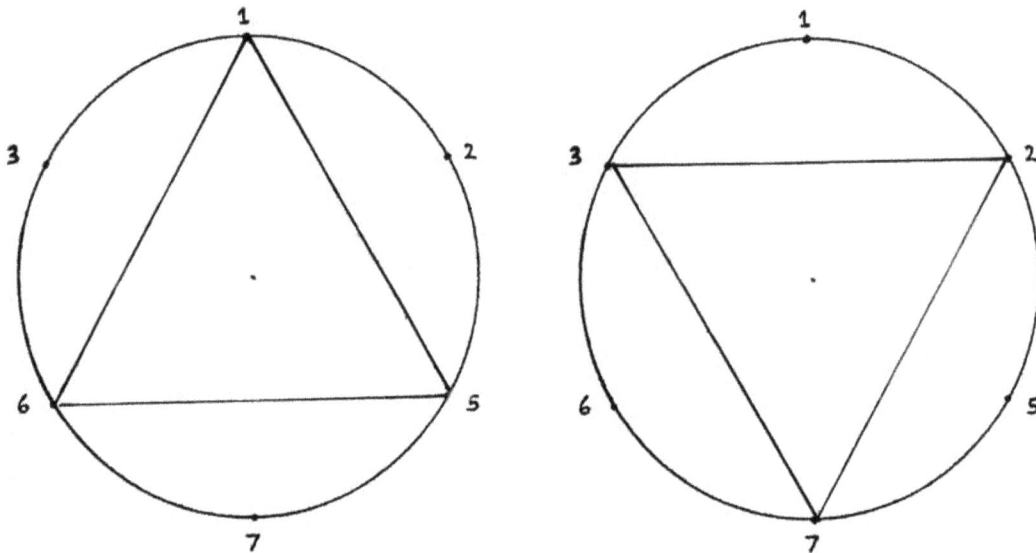

Fig 3b
The two triangles summing to 12,
one is upright, and one is reversed.

In esoteric lore, the Upright Triangle is often the Sun or Solar or Masculine energy. The shape can be seen as somewhat phallic, compared to the Downward pointing Triangle which is like the womb and therefore considered to be Lunar or of the Moon, and Feminine. Special symbols, in Alchemy, are attributed to these orientations.

●   The 2 large main triangles combined have sums of **24**. They form the traditional Star of David, sometimes known as The Seal of Solomon. The numbers include:
1 + 5 + 6 = 12
2 + 3 + 7 = 12

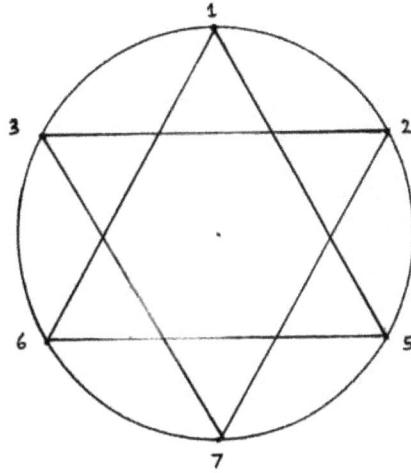

Fig 3c
Two Interlacing or Interpenetrating Triangles
whose combined sum is 24

● 3 Rectangles of 4 numbers sum to **16**. They include:

1 + 2 + 6 + 7 = 16 (shown in diagram Fib 3d)

2 + 3 + 5 + 6 = 16

1 + 3 + 5 + 7 = 16

This sum of 16 is double that of the Opposing Pairs.

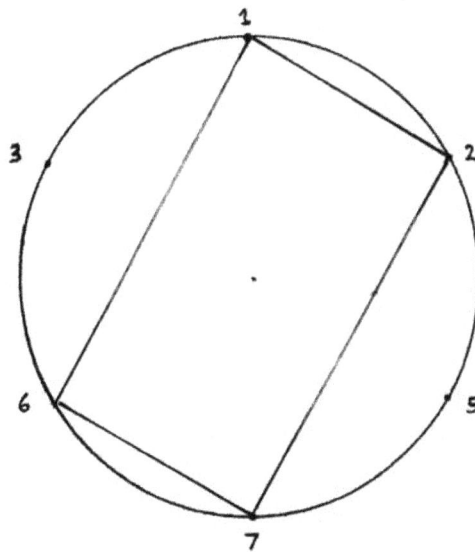

Fig 3d
1 of 3 possible rectangular array of 4 numbers that sum to 16

The interpenetrating male and female triangles (Star Of David shown in Fig 3c) was considered a fertility symbol. This concept of fertility can be seen in the Star of David crystal that is in a constant state of growth created when the whole shape is inserted into every 6th division or smaller circles. This process can go on forever, where a self-similar shape is repeated and repeated without destruction. It is called Fractal Geometry, the Mathematics of Infinity; a very exciting branch of mathematics that beckons more students to follow its course.

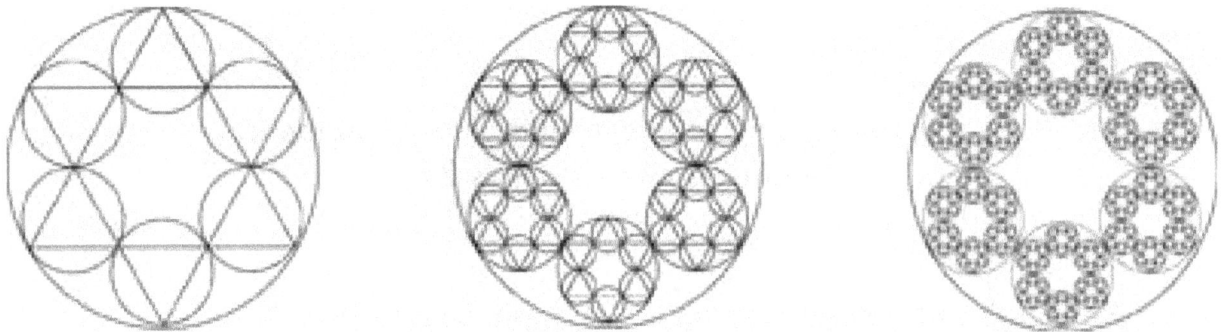

HOW TO CONVERT NUMBER INTO ART

The Process of converting Numbers Into Art, into Sacred Symbol will required the introduction of a basic circular net or Grid of 7 Dots in a circular arrangement as shown in Fig 4.

This design being created is not correctly known as a mandala, but rather as a "**Yantram**" which is the Sanskrit word for "Power Art", that which makes the Invisible, Visible. The Invisible is the **Mandala** (French: "Monde", as in world") or multi-dimensional view. eg: the Pentacle or 5 Pointed Star is the Yantram or 2-dimensional analogue of the 3-dimensional shape called the Dodecahedron, like a soccer ball, that has 12 distinct pentagonal faces. Here, we are drawing a shadow of a form, and our thoughts are carried in the psychic essences conducting through these lines, that is why it important what our intent is. Of more importance is the state of mind we are in when we are drawing this, so that we are focused and mindful.

Fig 4
The Grid of 7 Dots in a circular array necessary to convert Number Into Art

Fig 5
The Star Of David Yantram
Derived From The Numbers 1 To 7 drawn at 0°

Upon the 6 outer dots or division points, you can pencil in the 6 numbers (1,2,3,5,6,7) leaving out the centre number "4" which is implied. There is no need to write in the centre number, as it will interfere with the design construction.

By joining a long, unbroken or continuous line from 1 → 2 → 3 → 4 → 5 → 6 → 7, and joining the last to the beginning, the Alpha and Omega, from 7 to 1, a circuit is created. It does not look aesthetically beautiful yet, but it will evolve with 2 more superimposed rotations upon itself.

We have to give this Yantram a name, thus it will be known as: **"The Star Of David Derived From The Numbers 1 To 7 drawn at 0°"**

For convenience, we place the number "1" at the topmost north or zenith point of the circle.

The Yantram is created by Joining The Dots in Consecutive Order:

The order of the natural counting numbers: 1,2,3. This is basically a search for Order amidst apparent Chaos or Disorder.

You could perhaps search for more Order or Symmetry, by applying some origami or paper folding of the design, eg: assuming the main axis is the line from 3 to 5, and you fold the design upon itself, does it demonstrate mirror-imaging? so that one half of the image superimposes exactly upon the other half of the image. The two foldings along the two main axes are shown in Figs 5a, for experimental purposes.

The following 2 diagrams Figs 5a and 5b are merely **experimental designs** to see what happens when we fold the Yantra along its two longest lines running from "1 to 7" and "3 to 5". They are not exciting designs, but are used just for the record, to give us glimpses into the nature of the encircled Star of David's hidden geometrical surprises.

Fig 5a	Fig 5b
The Star Of David Yantram	**The Star Of David Yantram**
Derived From The Numbers 1 To 7	Derived From The Numbers 1 To 7
drawn at 0° and folded upon itself	drawn at 0° and folded upon itself
along the main long-lined axis "1 to 7"	along the main long-lined axis "3 to 5"

We now superimpose the same pattern upon itself, but at a 60° angle. The two self-similar designs convert the definition of the singular Yantram to its plural terminology of "Yantra". Thus it is known as:

"The Star Of David Yantram Derived From The Numbers 1 To 7 drawn at 0° + 60°".

To achieve this, draw Fig 5 first, say in one colour, then with another chosen colour, superimpose the second rotated pattern upon itself, by turning the numbers around, like a dial moved one notch in a clockwise direction, so that the zenith point is now number "2", as shown below. Reading clockwise, and from the top most point, the circular sequence of 6 numbers reads as: "2-1-3-6-7-5".

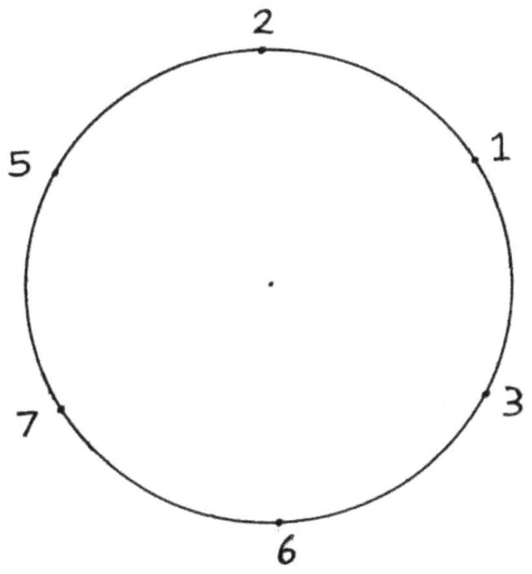

Fig 6
"The Star Of David Yantra
Derived From The Numbers 1 To 7"
has its wheel of numbers rotated
clockwise @ 60°
ready to be superimposed upon itself.

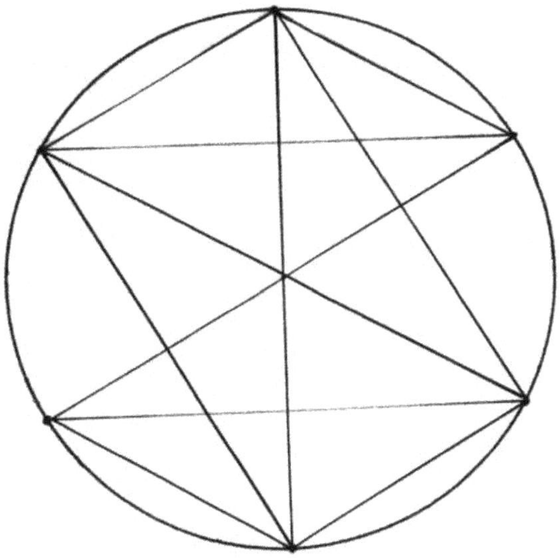

Fig 7
The Star Of David Yantra
Derived From The Numbers 1 To 7
drawn at 0° + 60°

To complete the sacred symbol pattern, we superimpose a 3rd self-similar design at 120° upon itself. We could use the jargon that we have drawn it 3 times at "0° + 60° + 120°" but this is too long-winded, thus we write it in its simplified form:

"The Star Of David Yantra Derived From The Numbers 1 To 7 drawn at 3 x 60°"

You may want to draw this 3rd pattern in another colour different to the previous two.

Reading clockwise, and from the top most point, the circular sequence of 6 numbers reads as: "5-2-1-3-6-7".

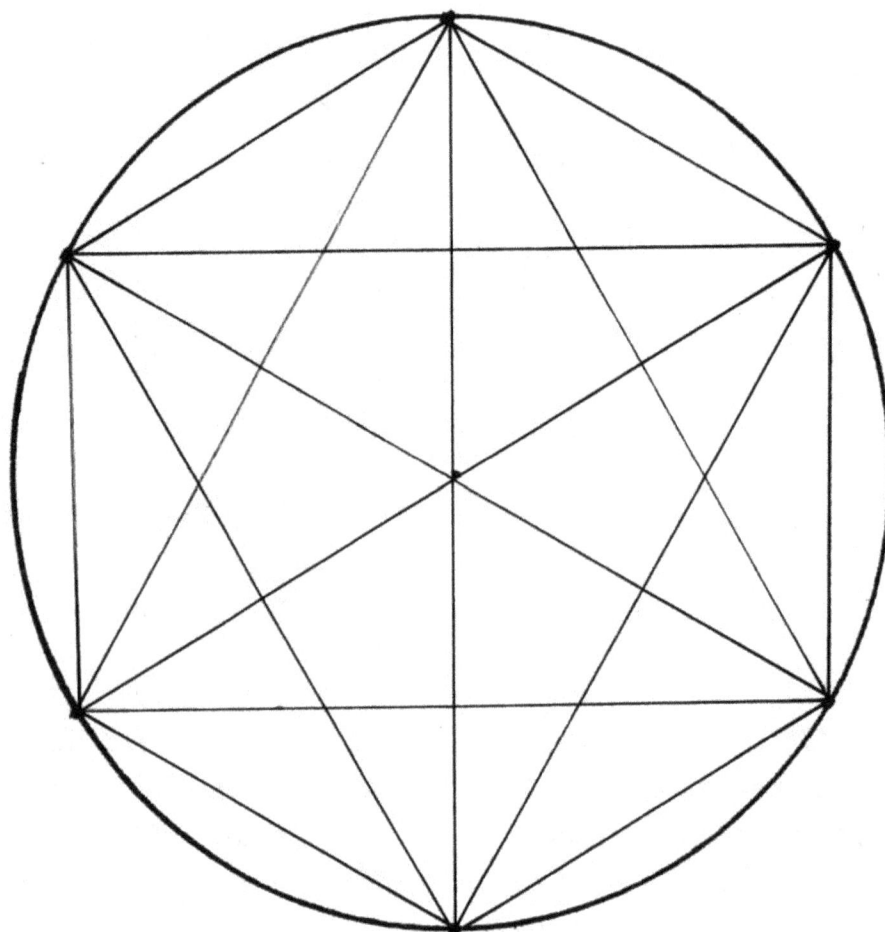

Fig 8
The Star Of David Yantra
Derived From The Numbers 1 To 7 drawn at 3 x 60°

Now the pattern is balanced and complete. This is the numerical and geometrical derivation of The Star Of David.

Thus if you read a cryptic line in an esoteric book, "...the 6 From the 7..." you could establish that the 6 Pointed Star of David was had it's mathematical derivation from the first 7 numbers.

It is a Sacred Symbol that contains both the Hexagram Star and the 6-sided Hexagon shape.

It has 3 geometrical components:

1 – The **Hexagon**, as the outer boundary
2 – The **Hexagram**, as the inner star
3 – The **Coptic Cross**, as the inner 3-dimensional form of the Christian Cross which is made up of the 3 internal axes seen as the 3 diagonal lines passing through the center. (The origin of the Coptic Church has its roots in Egypt). As a 3-dimensional cross, these 3 axes are at 90 degrees to one another:

I have redrawn and changed these diagrams below from poor internet images in an attempt to capture this 3-dimensionality inherent in the Star Of David. Really, the hexagon is the shadow of the Cube!

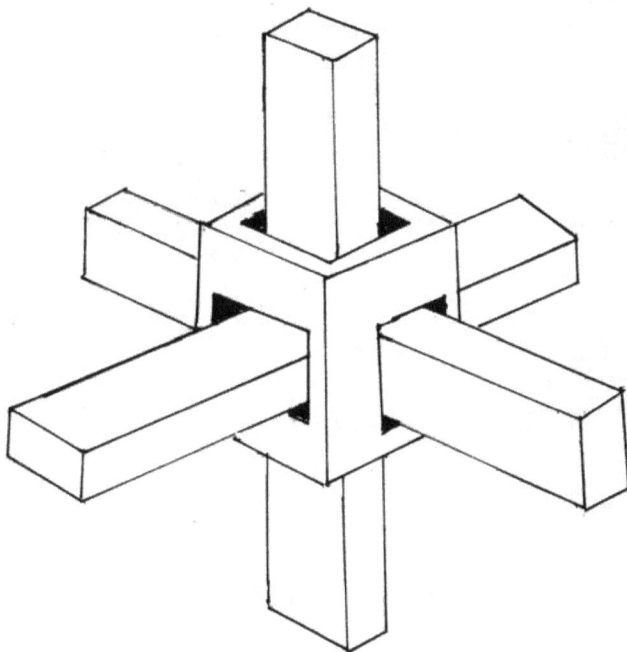

Fig 8a
Star of David composed of 60 degree angles, is perceived as 90 degree angles of the cube! This diagram is the inside of the Cube, yet the Cube, when tilted forms the Hexagon.

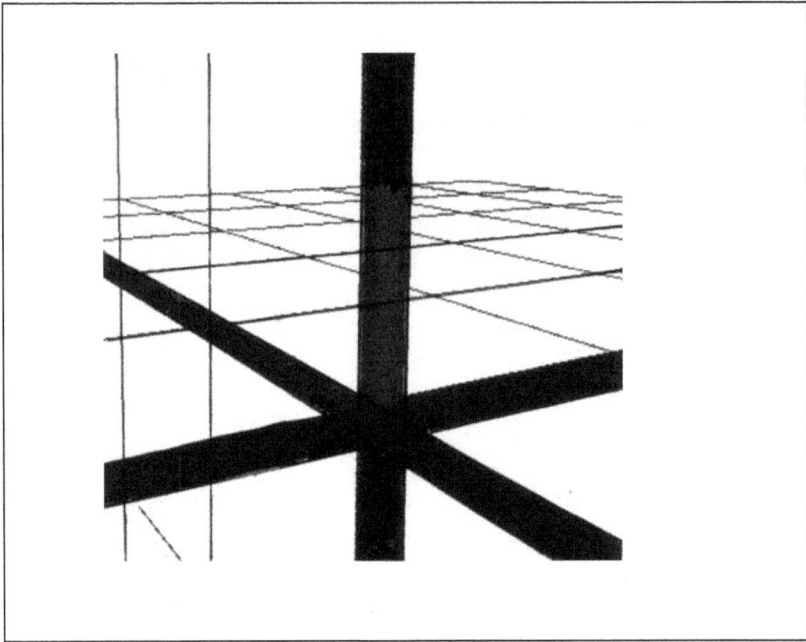

Fig 8b
The Coptic Cross has 3 axes that pass through the centre of the cube.
It's 3-dimensional analogue is the 2-dimensional Star of David?

This interplay of Hexagram Star and Hexagon Form, can be seen in the following two diagrams, Figs 9 and 10:

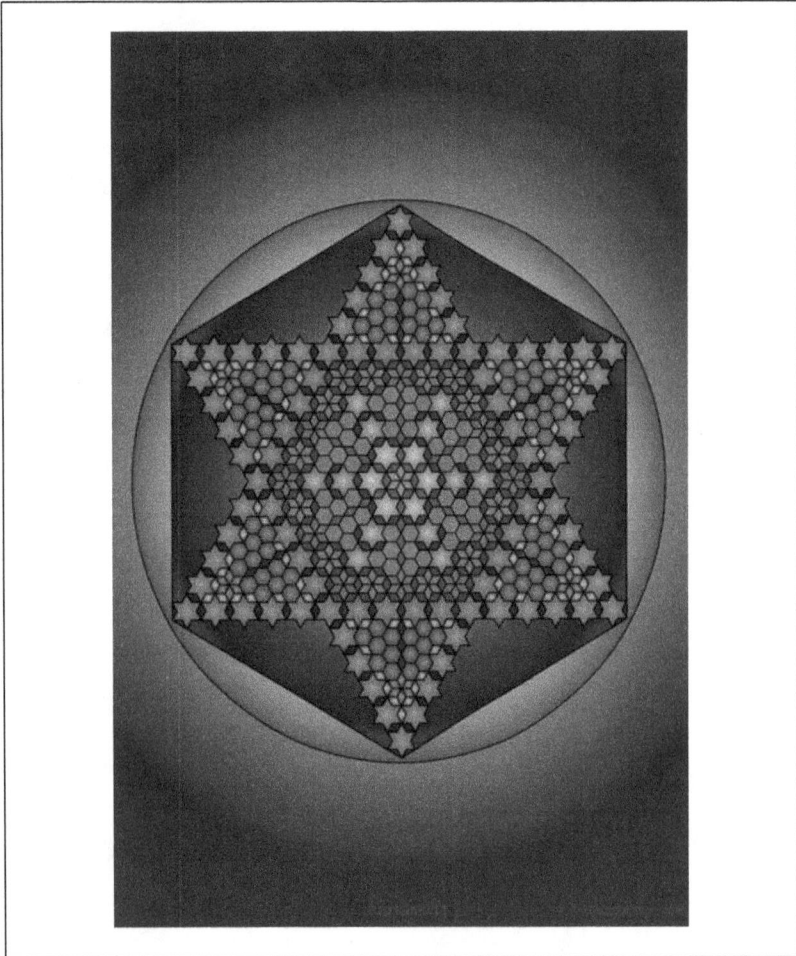

Fig 9
Star of David Design by **Jonathan Quintin**

(A4 prints of this an many others are available from www.sacredgeometry.com.au)

Fig 10
Islamic Tiling showing the negative space
as a traditional Star of David and Hexagonal gaps.

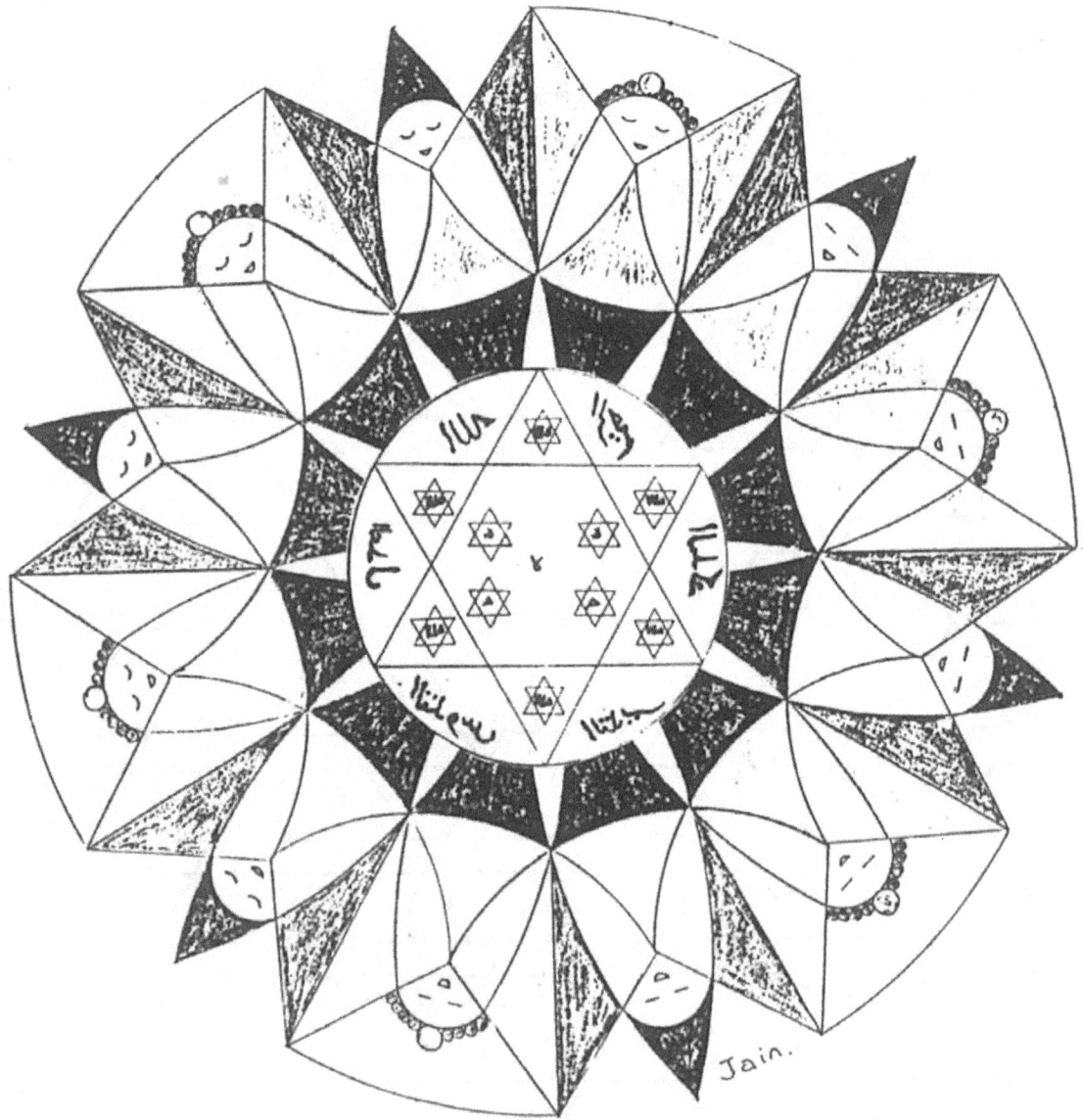

Fig 11
"Circle of 12 Beings In The Golden Ratio"
Drawing by Jain 1982
showing alternating Male and Female Forms.
Central Motif is the Star of David.

QUESTION: Can you find or detect any other Harmonics or Symmetries when viewing this Star as 6 smaller triangle?

nb: The centrepoint or central cell is the Number 4.

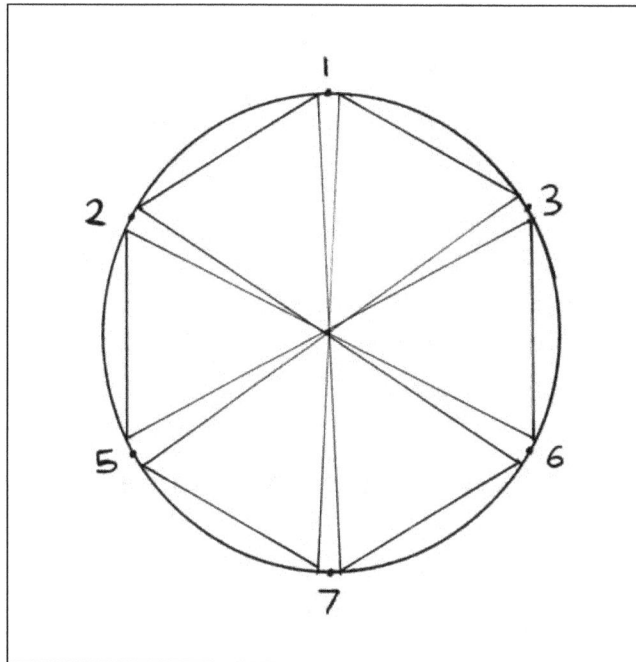

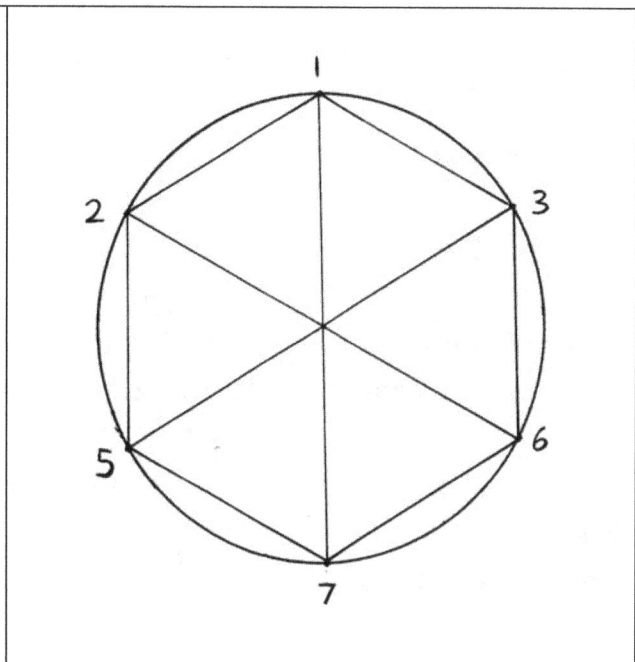

Fig 12a
The 6 distinct smaller triangles, Version 1, shown partitioned
to help add up the 3 corners easily.

Fig 12b
The 6 distinct smaller triangles, Version 2

This diagram of Fig 12a shows how the avid student Pattern Hunter divides the circle into 6 distinct smaller triangles, in her/her quest for further possible symmetry.

ANSWER see Fig 12b:
Here is the same diagram of the 6 smaller triangles.

Listing the 6 triangles, from top and clockwise, and adding their sums, as

Triangle 1 is   1+3+4 = 8

Triangle 2 is   3+4+6 = 13

Triangle 3 is   4+6+7 = 17

Triangle 4 is   4+5+7 = 16

Triangle 5 is   2+4+5 = 11

Triangle 6 is   1+2+4 = 7

We can see from the 6 sums that they are all various sums.
The sum of all the triangles is 72.

Though when you add the sum of 2 opposing triangles, their combined sum is **24**.
Here they are listed:

Triangle 1 + Triangle 4 = (1+3+4) = (4+5+7) = 24

Triangle 2 + Triangle 5 = (3+4+6) = (2+4+5) = 24

Triangle 3 + Triangle 6 = (4+6+7) = (1+2+4) = 24

We can see that the sum of the 3 opposing pairs is 3x24 or 72.
72 has many special properties. It is an anointed number, but will not be discussed here.

Let us draw these 3 opposing triangles, and add up their sums, but this time, without adding the central number "4" twice.

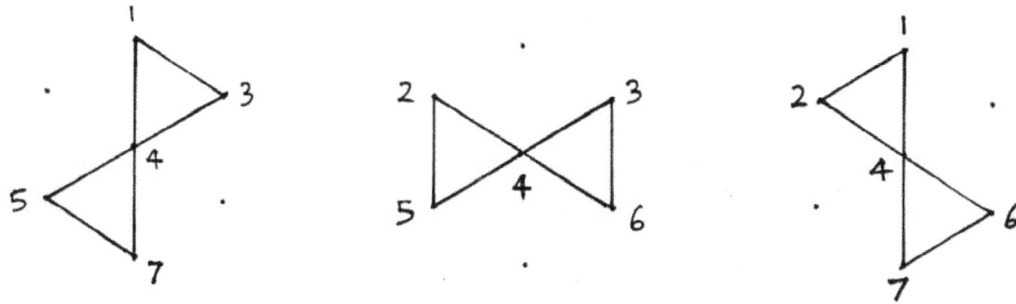

Fig 12c
The 3 Opposing Pairs of Triangles in the Star of David,
Whose sums all equal 20, 20, 20.

Adding the numbers of the 3 Opposing Pairs or Bowtie Forms, starting from right to left in Fig 12c gives:

$1+3+4+5+7 = 20$

$3+4+6+2+5 = 20$

$4+6+7+1+2 = 20$

Our conclusion is that there is more hidden symmetry when we connect and add the sums of the 3 possible bowtie forms. Each bowtie includes 5 numbers whose Constant or Magic Sum is 20. Thus we have a "**20, 20, 20 Code**"

At this point, the Teacher can ask the Student, if they know of any other Yantram that has 3 sums of 20.

Here is the **Solution**:
It can presented as: **DID YOU KNOW?**

This realization of "**20:20:20**" is cross-referenced to another cultural Mathematical symbol.

219

Did You Know that there is an ancient Indian Mathematical Triangle dedicated to the Goddess Ambaji. She is a Goddess of Creation, as the Universe flows through her breast. Her vehicle is the Tiger. At the base of her feet is the Magic Triangle, composed of the 9 numbers from 1 to 9, and arranged triangularly so that the 3 sides have equal sums of 20.

I first wrote about this in THE BOOK OF HARMONY SQUARES a large A3 size self-published in 1990, now an A4 size and renamed as THE BOOK OF MAGIC SQUARES, volume 1 (of a 3 part series).

3 SHREE AMBAJI

Fig 13a
Goddess Ambaji, form of Durga, riding her Tiger. Inscribed near her feet is the Magic Triangle Yantra with Sanskrit numerals composed of the first 9 consecutive numbers whose 3 sides sum to 20.

Inside the illustration:

from India.

Triangle numbers:
1, 6, 7, 9, 5, 8, 4, 3, 2

Mata Amboji's symbol of the Majik TriΔngle whose summations on three sides = 20. The weapons in her hands indicate power over negative mental states e.g: the Sword cuts thru ignorance, the Axe roots outs greed, the trident pierces thru illusion. etc.

Fig 13b
The Magic Triangle Yantra with Sanskrit numerals composed of
the first 9 consecutive numbers whose 3 sides sum to 20.
The border, by Jain, has been drawn with the Magic Square of 3 Yantra
pattern, at 0 degrees, tiled side by side at right angles to one another.

QUESTION:
Can you look or squint your eyes again at the The Star Of David Yantra Derived From The Numbers 1 To 7, and find another harmonic pattern by isolating 3 small triangles. What is the Magic Sum?

ANSWER:

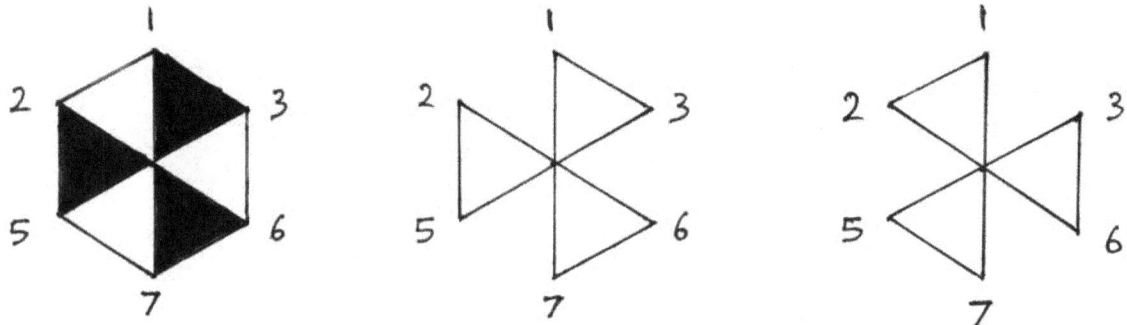

Fig 14
The sum of 3 Triangles

There are Two Methods How to Add the Sum of the Triangles.
Method 1 is include the centre no. "4" in each count of each triangle; and Method 2 does not use the centre no. "4" in each count of each triangle. This will result in two differing Magic Sums.

METHOD 1:

The Shaded Area of the 3 small Triangles in Fig 14 is:
(1+3+4) + (4+6+7) + (2+4+5) = 36.

The Unshaded Area of the 3 small Triangles in Fig 11f is:
(1+2+4) + (3+4+6) + (4+5+7) = 36.

We notice here that both sums of Shaded or Unshaded triangular areas are **36**.

METHOD 2:

This time we do not add the centre number "4" when adding the sum of 3 triangles, we only add it once. Instead of adding 9 numbers (3x3), we are only adding 7.

The Shaded Area of the 3 small Triangles in Fig 14 is:
1+3+4+6+7+5+2 = **28**.

The Unshaded Area of the 3 small Triangles in Fig 14 is:
1+2+4+3+6+7+5 = **28**.

We notice here that both sums of Shaded or Unshaded triangular areas are **28**.

Fig 15
The Organic form of the Star of David
taken from l'Archeometry
by **Saint Yves D'Alveydre**

The UNICURSAL HEXAGRAM

QUESTION:

Out of curiosity, is it possible to draw a variant of this stylized Star of David, with a circle around it, in another form, where the 6 points on the circle are connected in such way, with straight lines only. The only condition is that you can not take your pen or pencil off the paper, a bit like the familiar pentacle or 5 pointed star that most children can draw very easily. Here is the image of Fig 4 below for you to practice on.

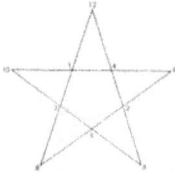

Unicursal literally means "One-Line".

Nb: The Hexagon is one solution, but does not count, as we want a starlike form or Hexagram, that has distinct intercepting lines.

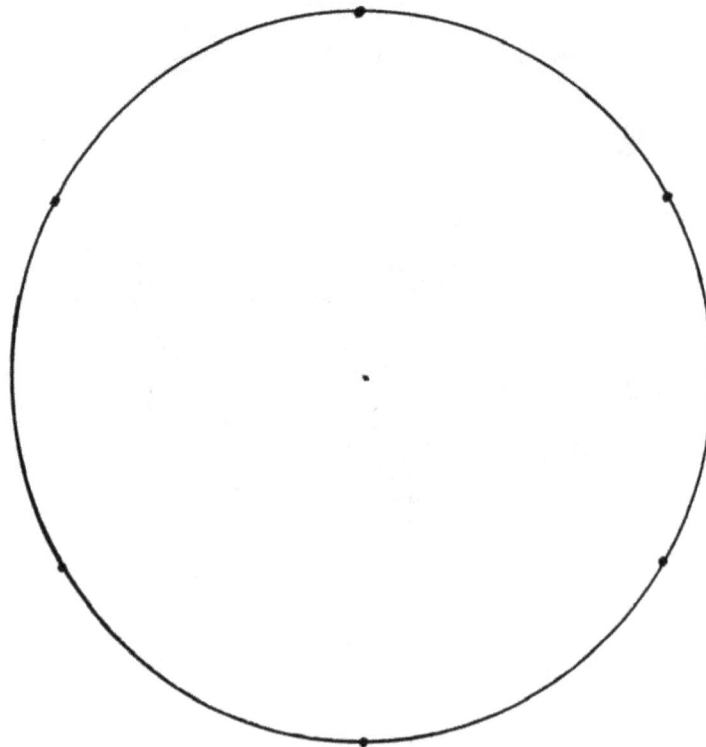

Fig 4

Circular Grid of 6 dots used to experiment with connecting them all without taking your pen or pencil off the paper.

ANSWER:

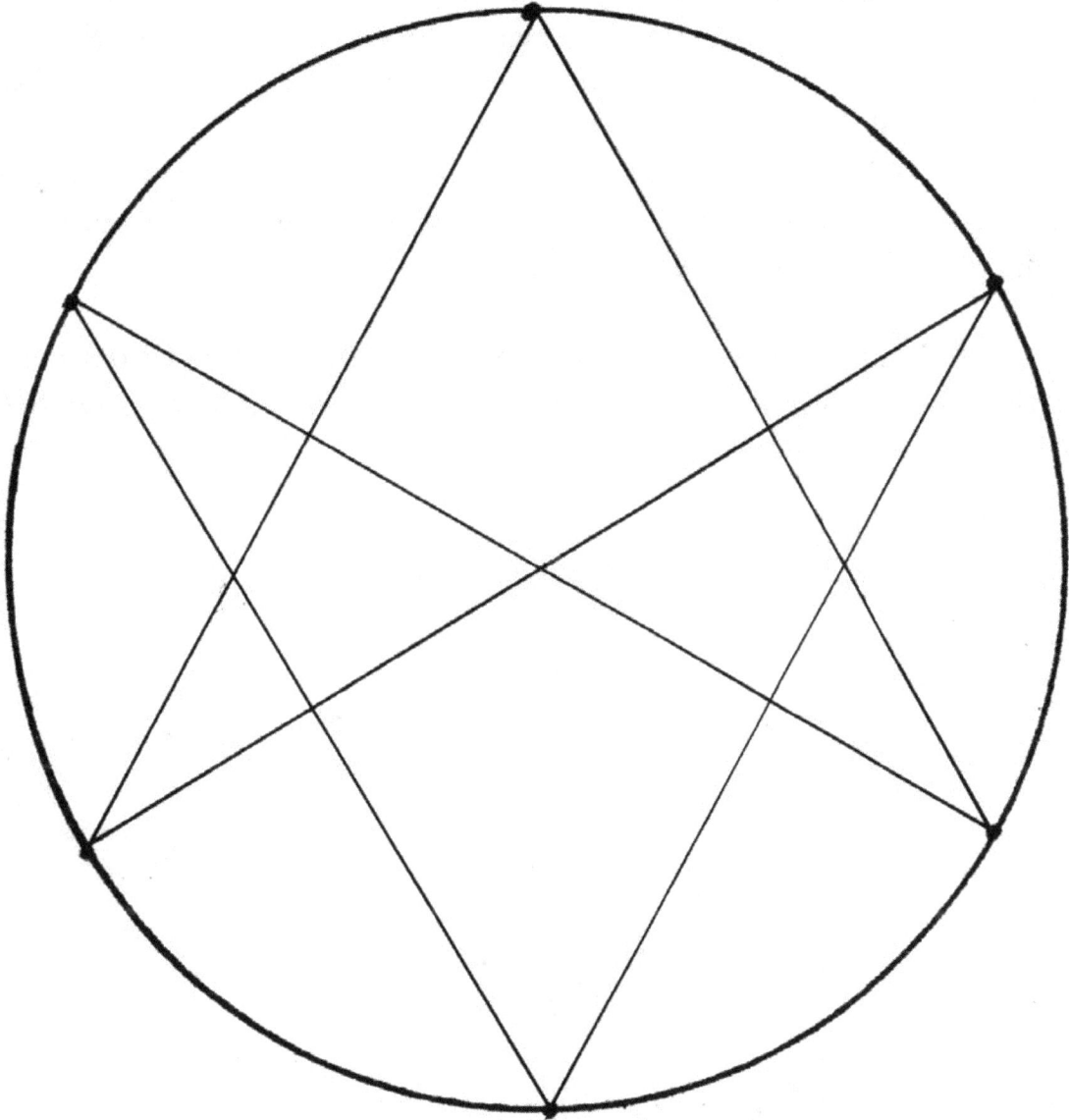

Fig 16
The Unicursal Hexagram, a variant form of the Star of David,
but is created without lifting the pen off the paper.

The solution is simple and beautiful especially when the long continuous unbroken line is seen as an electrified copper circuit whole electrons flow harmoniously forever in a closed loop.

Some occult magic orders, like the OTO of Aleister Crowley, revered this symbol.

It is quite amazing that this whole chapter of rare information was wielded out of the composition of just 7 numbers arranged into a simple star formation. It indicates that these sacred shapes generated from these numbers, are windows into the universe of creation. It is very similar to the 3x3 square of 9 numbers, the magic square of 3, that also creates volumes of fantastic designs, like the atomic structure of diamond lattice etc...

This technique of the Art of Number is therefore important to pass onto future generations, for we currently only have a glimpse of the enormity of forthcoming symmetries yet to be revealed.

These next few pages show 3 pages of extracts (Figs 17a,b,c) from my early hand-written books this one, still available, is known as: THE BOOK OF MAGIC SQUARES, volume 3 published 2000.

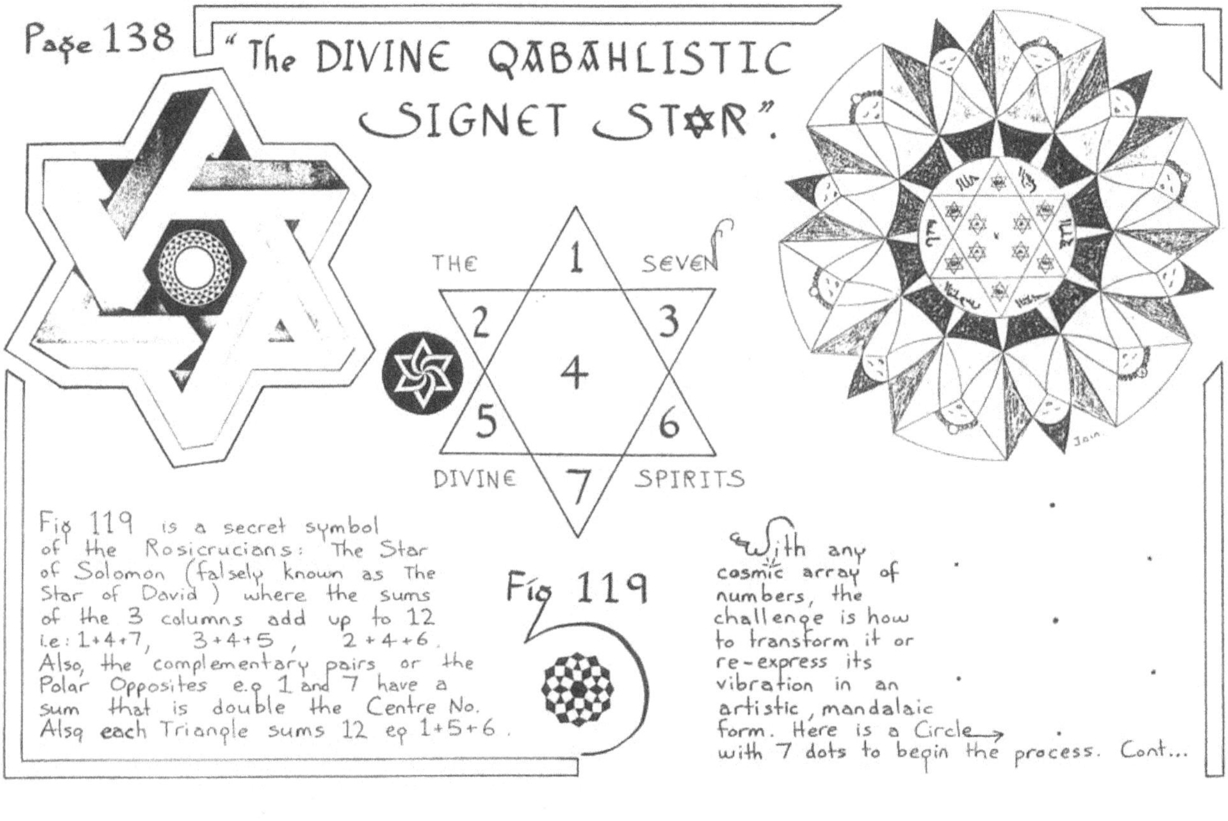

Fig 17a

"The Star of David Derived From The Numbers 1 to 7" in its hand-written form circa 1999. I coined the name "The Divine Qabahlistic Signet Star".

Fig 17b

"The Star of David Derived From The Numbers 1 to 7" in its hand-written form circa 1999. The second page which allowed the reader to draw the geometries.

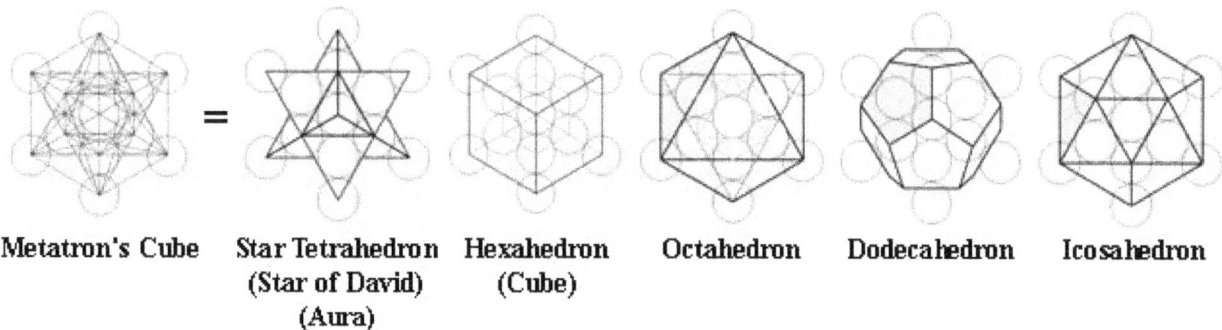

| Metatron's Cube | Star Tetrahedron (Star of David) (Aura) | Hexahedron (Cube) | Octahedron | Dodecahedron | Icosahedron |

Fig 17

Other variations of 6 pointed-ness, known as the **"Fruit Of Life"** having 13 faint circles in a hexagonal configuration, revered by the Ancient Seers, as this grid was capable of deriving the 5 known Platonic Solids: the key to all Atomic Structure, one shape or more being embedded with another like Russian Dolls.

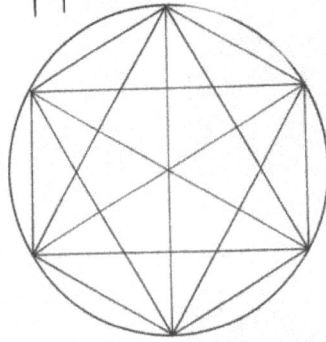

fig 121

Magic Hexagon

fig 120

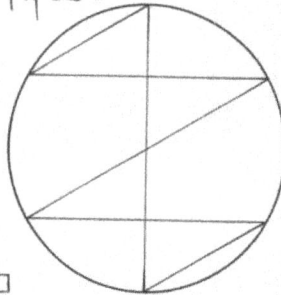

Magic Cross

12	9	
2	7	6	11
5	3	10	8
4	1	

Fig 17c

"The Star of David Derived From The Numbers 1 to 7"
in its hand-written form circa 1999.
The 3rd page showing the solutions to the puzzle, at the end of the book.

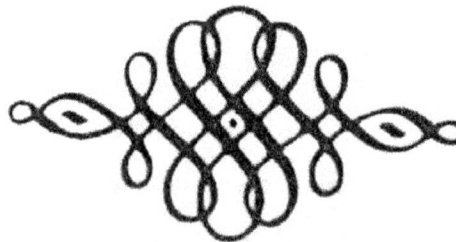

228

CHAPTER 8

THE FLOWER OF LIFE
GENERATED FROM THE GEOMETRY OF THE
STAR TETRAHEDRON

CHAPTER SUMMARY:

- The Flower of Life Pattern laser-carved at Osirian Temple in Abydos
- Tetrahedron and the 4-Frequency Tetra
- 64 Tetrahedra Crystalline Matrix
- Star Tetrahedron
- Cuboctahedron (aka Vector Equilibrium)
- Foo Lions of the Forbidden City have 3-Dim Flower of Life Under Paw
- Hebrew Qabahlistic 10-Sphered Tree of Life Versus the Original 12-Sphered Tree of Life
- 12 Sphered Tree Of Life Forming the Macrocosmic Snowflake
- Chinese I-Ching Trigrams can be rearranged to form the Star Tetra

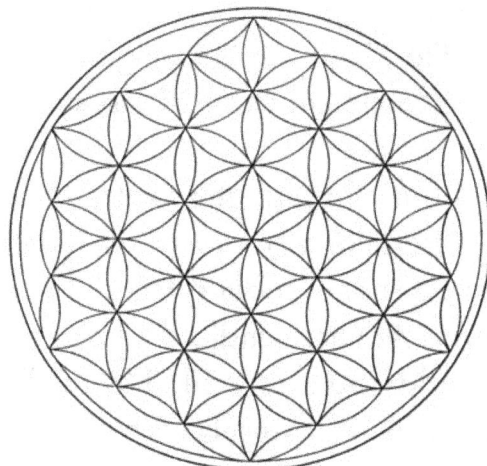

'The Flower Of Life'

The Flower of Life Pattern

Fig 1
Traditional Flower Of Life pattern is the blueprint or template of all creation.
(image computerized by Jonathan Quintin: www.sacredgeometry.com)

This pattern has been found all over the world, thousands of years ago. It is seen in it's 2-dim iconic form high in the ceiling of the Egyptian or Osirian Temple of Abydos on the Nile River, yet the enigma is, it was not carved, it was mysteriously burnt into the atomic structure of the stone, as if by a laser beam, thousands of years ago!

Every circle can be seen to pass through the centre of a neighbouring circle; the union of such two circles is traditionally known as the Vesica Piscis. Children, as young as 5 years old, can be seen doodling this pattern of intersecting circles to derive a similar flower form. Adults, mindlessly talking on the telephone, whilst doodling, invariably draw this petal shape. It's as if it is fully known in the depths of the mass consciousness. Merely drawing it, by hand or compass, brings a sense of peacefulness or wholeness.

In this chapter we will explore how this Flower of Life pattern can be generated from various primal shapes. We begin with another fundamental form, the Tetrahedron, which is like a triangular based pyramid. The tetrahedron is the basic building block of all atomic structure.

Let us look at subdivisions of the base and sides. If we explore midpoints of all 6 edges, we would have a 2-frequency tetrahedron, essentially composed of 8 smaller tetrahedra (since $2^3=8$). If we subdivided the 6 edges into three equal parts we would have a 3-frequency tetrahedron, essentially composed of 27 smaller tetrahedra (since $3^3=27$). But here we are interested in a 4-Frequency Tetrahedron essentially composed of 64 smaller tetrahedra (since $4^3=64$).

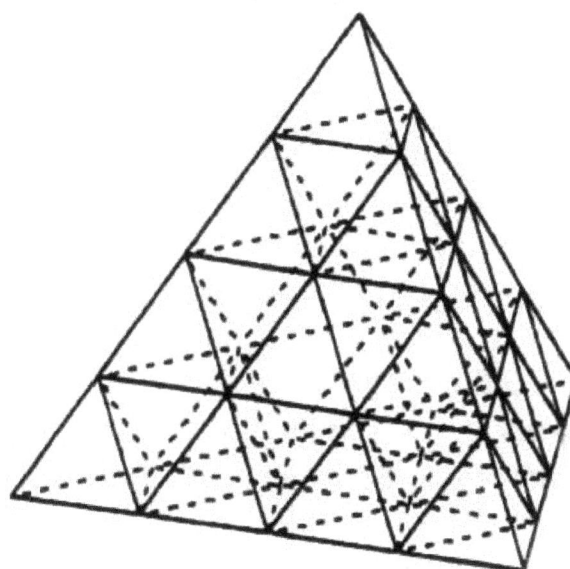

Fig 2
A 4-Frequency Tetrahedron
(the original triangular base is divided into 4 parts)

From a mathematical perspective, or number theory, it's interesting that this well known sequence of cubic numbers: 8, 27, 64, 125, 216, 343, 512, 729, etc are obviously known to form cubes, but not obviously known that the shape can be rearranged to form tetrahedra (the plural of "tetrahedron").

This tetrahedron interpenetrates itself to form the Star Tetrahedron. We have already seen in a previous chapter how the 2-dimensional shadow of this form, being the Star of David, can be formed in two distinct ways:
1 - from the use of numbers from 1 to 7 and
2 - using Row-By Row Analysis of the Magic Square of 3.

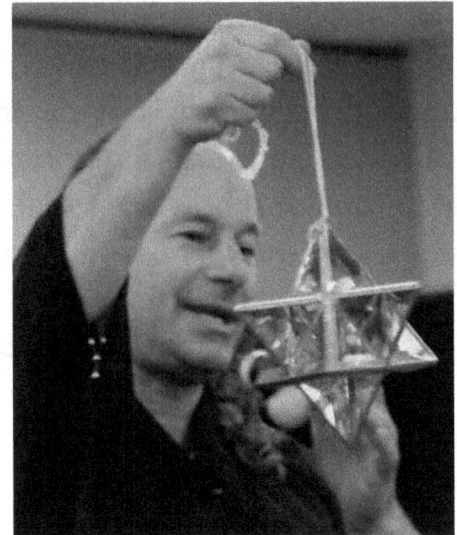

Fig 3
Star Tetra models
(from Leonardo da Vinci, wooden model + Jain holding a glass Star Tetra
made passionately by Asaf Zakay in Byron Bay
www.zakayglasscreations.com).

The Star Tetrahedron's 3-Dim form is really a cube (when you join the 8 vertices). So really whenever you see the flat form of the Star of David and the shape around it called the Hexagon, you are really perceiving the Cube.

Interestingly, at the core or centre of this crystalline construct of 64 Tetrahedra is the Cuboctahedron (aka Vector Equilibrium) that has 12 perfectly balanced inner radials or vectors. It's an ancient puzzle. If I had an orange, and I asked you how many other oranges, of the same size, can fit or pack or nest perfectly around this one central sphere, what would the magic number be. (In 2-dimension, the same puzzle would be: how many 20 cent coins can fit around a central coin, the answer being 6 exactly).

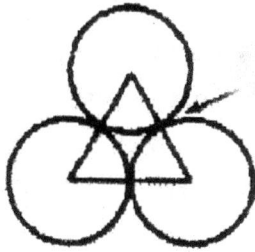

Fig 4a
0-Frequency Triangle
(1 Interval = 1 Edge)

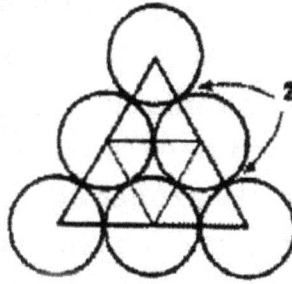

Fig 4b
2-Frequency Triangle
(2 Intervals = 2 Edges)

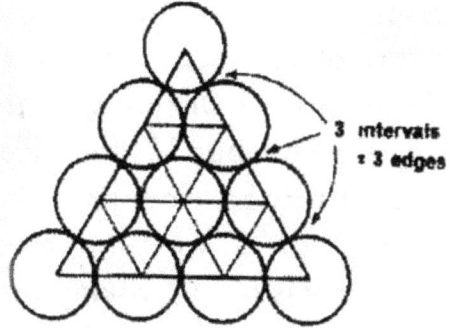

Fig 4c
3-Frequency Triangle
(3 Intervals = 3 Edges)

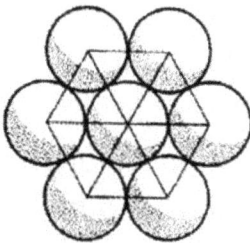

Fig 4d
Closest Packed Sphere
In The Plane

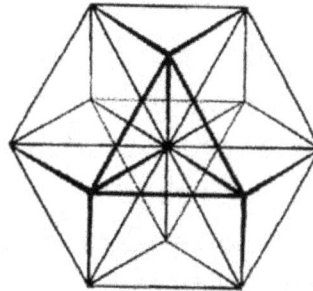

Fig 4e
Vector Equilibrium

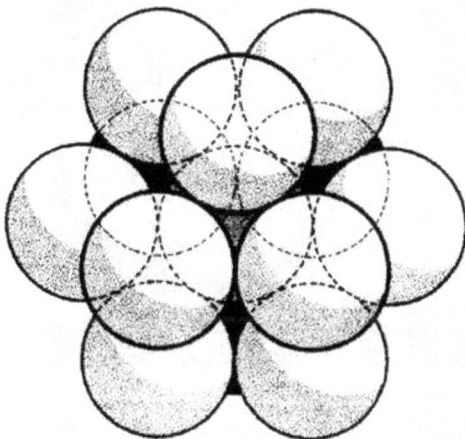

Fig 4f
Omni-Directional Closest Packed Spheres
Aka Vector Equilibrium

(This TWELEVE AROUND THE ONE
is really another expression of
the 13 CODE)

Fig 4
Cuboctahedron or the Twelve Around The One, is formed naturally from
the 12 spheres packing perfectly around a 13th or central sphere.

Essentially, the Cuboctahedron of 6 squares and 8 triangles is formed from the centre of 12 spheres that surround a central 13th likened to Jesus magnetizing to himself the 12 Disciples, constituting a veritable 12:13 Code). This is one reason why the Number 13, for millennia, has been instilled with fear, so that you don't open it's door. How ridiculous that skyscrapers omit the numbering of the 13th floor! That is why the relearning or remembering of this sacred (scared) geometry is about waking up.

The **"Twelve Around The One"** also known as the principle of "sphere kissing" that determines all atomic structure) as legendary Buckminster Fuller called it, is a symbol therefore that unites the world, like the torus doughnut ring, like communism or shareability, in stark contrast to the political symbol of the pyramid where the millions of debt slaves are at the base serving the elite plutocracy or handful of families that secretly govern the world.

The purpose for giving you this information was to determine how does the Flower of Life pattern emerge from these primal shapes?

By placing a sphere around each of the 64 Tetrahedra then rotating this construct, as if to view it's shadow, which is really the transduction of a higher 3-dim form down to it's lower 2-dim flatland form, we see emerging the distinctly recognizable Flower of Life form.

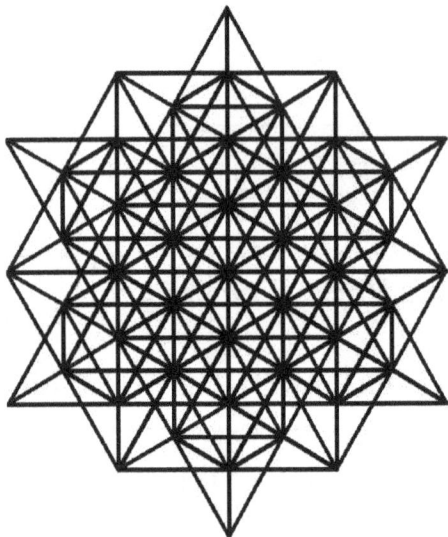

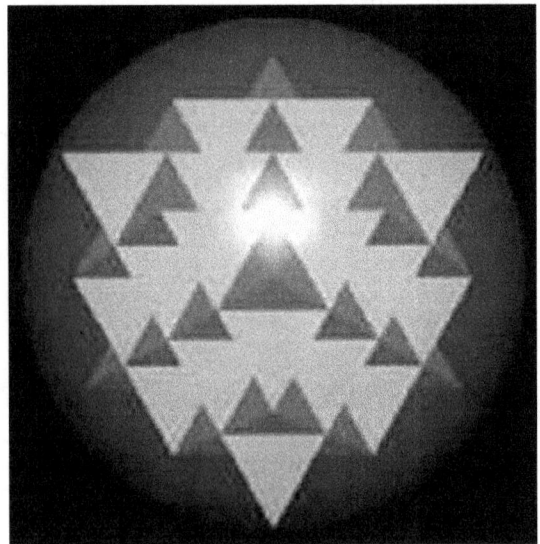

Fig 5

The 64 Tetrahedra, when visualized within 64 spheres
is a multi-dimensional form, that when tilted at a critical angle,
reveals the shadow of the familiar Flower of Life pattern.

234

In China, we see this Flower of Life pattern again, but in it's 3-dim spherical form. Around 1420, in the Forbidden City where the Sun Gods reside, are the large statues of the Foo Lions at the entrance of the Temple in Beijing. They act as Guardians of the Knowledge. Literally, the globe of the Flower of Life is under it's powerful and aggressively protective paw!

Fig 6
The fierce Foo Lions, protecting under it's right paw,
the emblem of the 3-Dimensional Flower of Life Sphere,
as if it were a cosmic treasure.

This Flower of Life or blueprint of creation, appears in other cultures:

1 - In the Hebrew Qabahlistic Tree of Life (all 5 images of Fig 7) showing in video animation style how 6 Tree of Life diagrams come together hexagonally to form the Flower of Life pattern yet again. . It can be seen clearly in a video animation by Nassim Haramein of the following geometry (visit: www.theresonanceproject.com).
2 - In the Chinese culture, hidden in the oracle of the "I-Ching" or "Book of Changes".

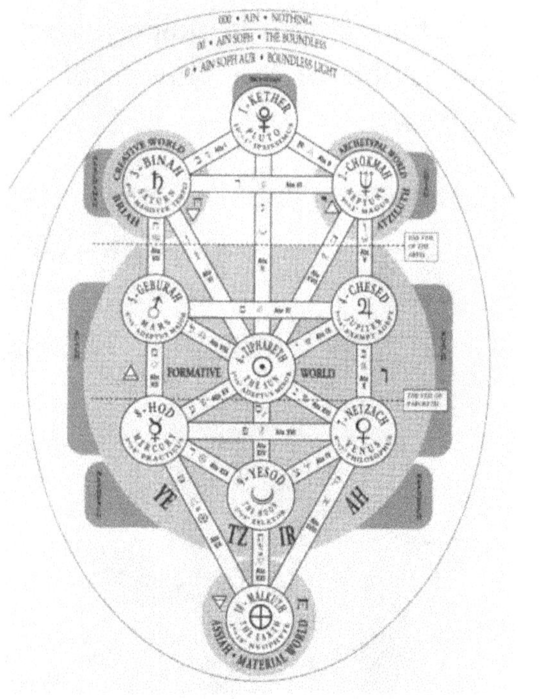

Fig 7a
The standard Qabahlistic Tree of Life showing the 10 Sephiroths or Spheres or Planetary Globes, showing a basic pathway from No.10 Malkuth (Earth) to No.1 Kether (Heaven) and the 22 pathways in between on Life's Journey back to the Source. (The 22 spokes also represent the 22 Hebrew Flame Letters of their Alphabet).

Fig 7b
6 hexagonally arrayed "Tree Of Life" meeting at a common centre. Also known as the Macrocosmic Snowflake aka Tree Star, this esoteric pattern also reveals the Flower of Life at it core, and is indicative of the atomic structure, ancient brahmanic or priestly knowledge that was guarded and held secret from the masses. It's nesting form also hints at "fractality" where the "Inside is the same as the Outside".

Fig 7
The standard Qabalistic Tree of Life
and it's deeper meaning hidden in the MacroCosmic Snowflake

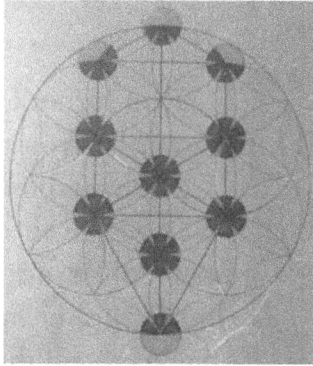

Fig 7c

The Tree of Life has been highlighted so that is visible in the Flower of Life.

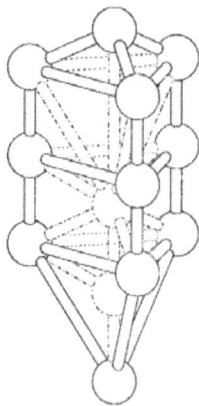

Fig 7d

Metatron (Lord Of The Electron) is the Guardian of the Tree of Life, as this geometry creates all the 5 Platonic Solids. It is shown here in it's crystalline or molecular form. Visualize the Macrocosmic Snowflake in Fig 7b in this multi-dimensional form.

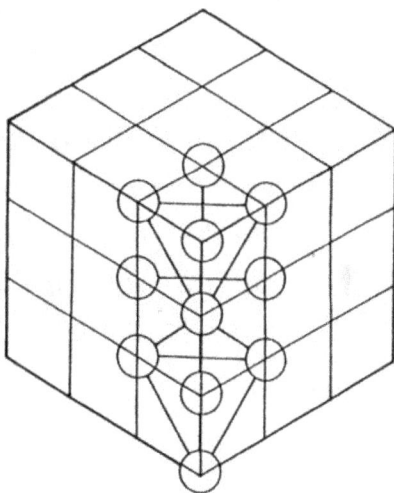

Fig 7e

The Tree of Life represented in yet another 3-Dimensional form, this time etched upon the edge of the Cube.

The sephiroth or sphere in the place of the very corner of the cube is traditionally known as the Abyss, and is not a planetary sphere, but the eye of the torus, the physics of the implosive Black Hole.

Fig 7
The Tree of Life visible in the Flower of Life +
3-D form seen in the edge of the Cube

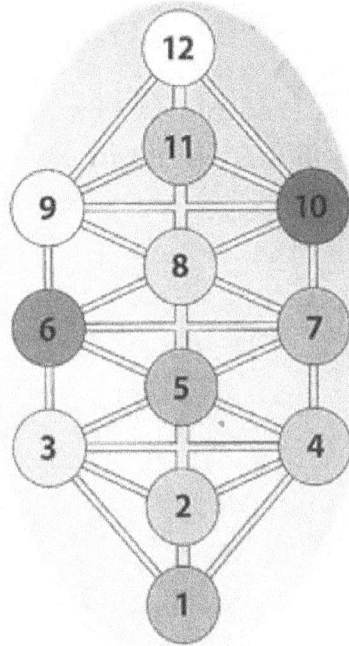

Fig 8a

True form of the Qabahlistic Tree of Life, is 12 Sphered, not 10. Our duty is to do the essential Research and apply Higher Intuitive abilities to sift the Chaff from the Grain.

(The graphic of the 12 Spheres "Kathara Grid" above is taken from Ashayana Deane's cosmological works).

Fig 8b

When the 12 Sphered Tree of Life map is overlaid correctly, via perfect fractal nesting or embedding, they form the Macrocosmic Snowflake, a signature for the Cosmos that contains all essential Penta-Hexa and the primal Square Root Harmonics.

Fig 8

The true form of the original Tree of Life, according to my research, is based on 12 Spheres or Sephiroth, not 10, as it forms a doubly terminated quartz crystal configuration that snugly fits the Base 24 Stargate Knowledge that literally "Opens The Door".

2 - The Flower of Life diagram is also seen in the Chinese I-Ching (oracle or divination technique used in The Book of Changes) composed of 64 Hexagrams. (A Hexagram is a glyph composed from 6 lines, where only Full Lines (Heaven) or Broken Lines (Earth) and a combination of these full and broken lines, are used to suggest a binary format).

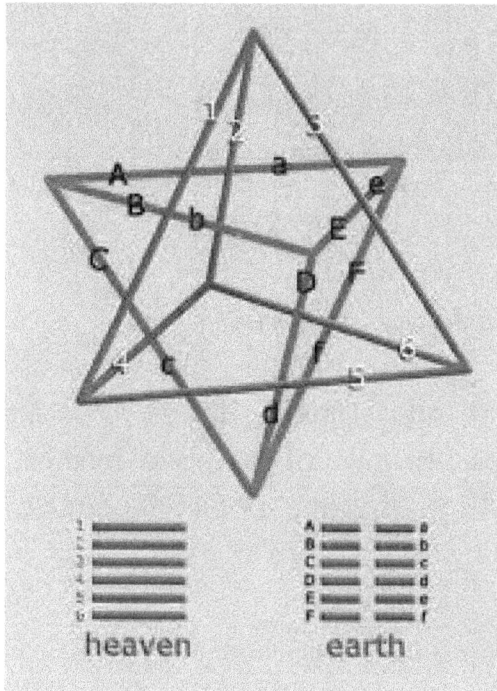

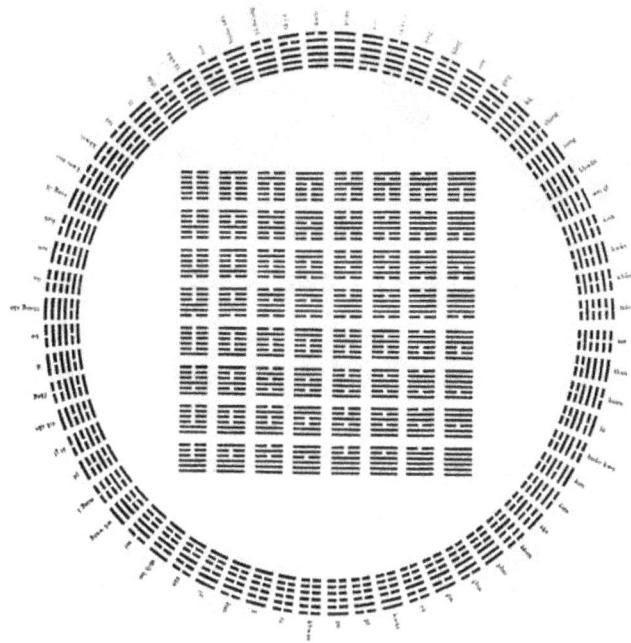

heaven   earth

Shao Yung's Old Family Mandala

Fig 9

The Flower of Life yantram is also seen in the Chinese I-Ching (oracle or divination technique used in The Book of Changes) composed of 64 Hexagrams as seen circularized in Shao Yung's "Old Family Mandala".

Using video animation style (by Nassim Haramein) to magically arrange these broken and unbroken lines, what emerges is the Star Tetrahedron, which is the fundamental building block of the Flower of Line pattern.

(In this animation, the 6 Full Lines first generate the Tetrahedron, as a Tetrahedron has 6 edges, and the remaining Broken Lines form around the Tetrahedron to magically produce the Star tetrahedron).

CHAPTER 9

SOUL SIGNATURES
BY JAIN 108

**Personalized Geometric SOUL YANTRA or Sigils
of your Name and Birthdate
derived from unique Magic Squares of 6x6.**

Have your Name and Birthdate converted into Art, all geometry hand-drawn by Jain, and artistically coloured by Lily Moses, special enough to have framed. Your Name and Birth-date are plugged into a Magic Square of 6x6 (a special Sun Code whose sums of rows, columns and diagonals have a magic sum of 111).

SOUL
SIGNATURES
100% Hand-Drawn by
LILY MOSES & JAIN 108

HAVE YOUR OWN GEOMETRIC SOUL SIGNATURE
CREATED JUST FOR YOU

This Geometric Soul Signature is constructed using a specific Magic Square Matrix of 6x6 Grid associated to the Sun whose Magic Sum equals 111. This "111 Unity Code" is the basis for your uniquely generated Yantra (Power Art) or Sun-Seal that has been carefully programmed Alpha-Numerically with letters of your Name and numbers of your Birth-date.

These sacred amulets or sigils were used and worn in ancient times to enhance personal well-being, protection, prosperity, love and harmony in one's life.

Your unique Geometric Shield is protected with a ring of Metatronic Script. This outer Ring acts as an amplifier of the sacred geometries encoded within.

Lovingly hand-drawn by 2 unique Byron Bay Vril-artists. Geometry by Jain 108 (www.jainmathemagics.com) and rendering of colour and infusion of celestial script by Lily Moses (www.lilymoses.com).

Place this Sun Code upon your altar or in a power spot in your personal space.

Your Soul Signature has been created from a carefully selected magic Square of 6x6 (A Galactic Sun Code) of which there are millions of variations, each one unique like a thumbprint or snowflake. Your name and birth-date have been literally plugged into a highly sophisticated and unique mathematical and geometrical matrix that has highly intelligent internal properties.

The Magic Square of 6 (111 Unity Consciousness Code) has all its columns, rows and diagonals summing to 111, a rare display of three ones or "thrice repeated unity" that constitute a Trinity, sacred to all cultures. Being purely based on the Universal Language of Mathematics, it therefore goes beyond all Eastern Vedic and Western cultures.

Author and mystic Idries Shah, in his book "The Sufis" wrote about this Magic Square of 6: Transposed into letters, 111 = "QUTUB", in Arabic. This is the reputed invisible head of all the Sufis. The word 'Qutub' literally means the Magnetic Pole, Pivot, Polestar, Chief. If 111 is split into 100, 10, 1 and substituted, we get the letters Q, Y & A. The word "QYAA" means "To be Vacant, Voided". It is this vacant, voided, purged "house" into which the "Baraka" (or the divine consciousness) descends.

Here are 3 examples of Geometric Soul Signature, the names on the yantra have been removed for protection of personal details, but this is enough to give you an idea how your letters of your Name and the numbers of your Birth-date can be translated into Art.

Cost is $300 to have one done, on A4 size art-paper, and $400 for A3 size.

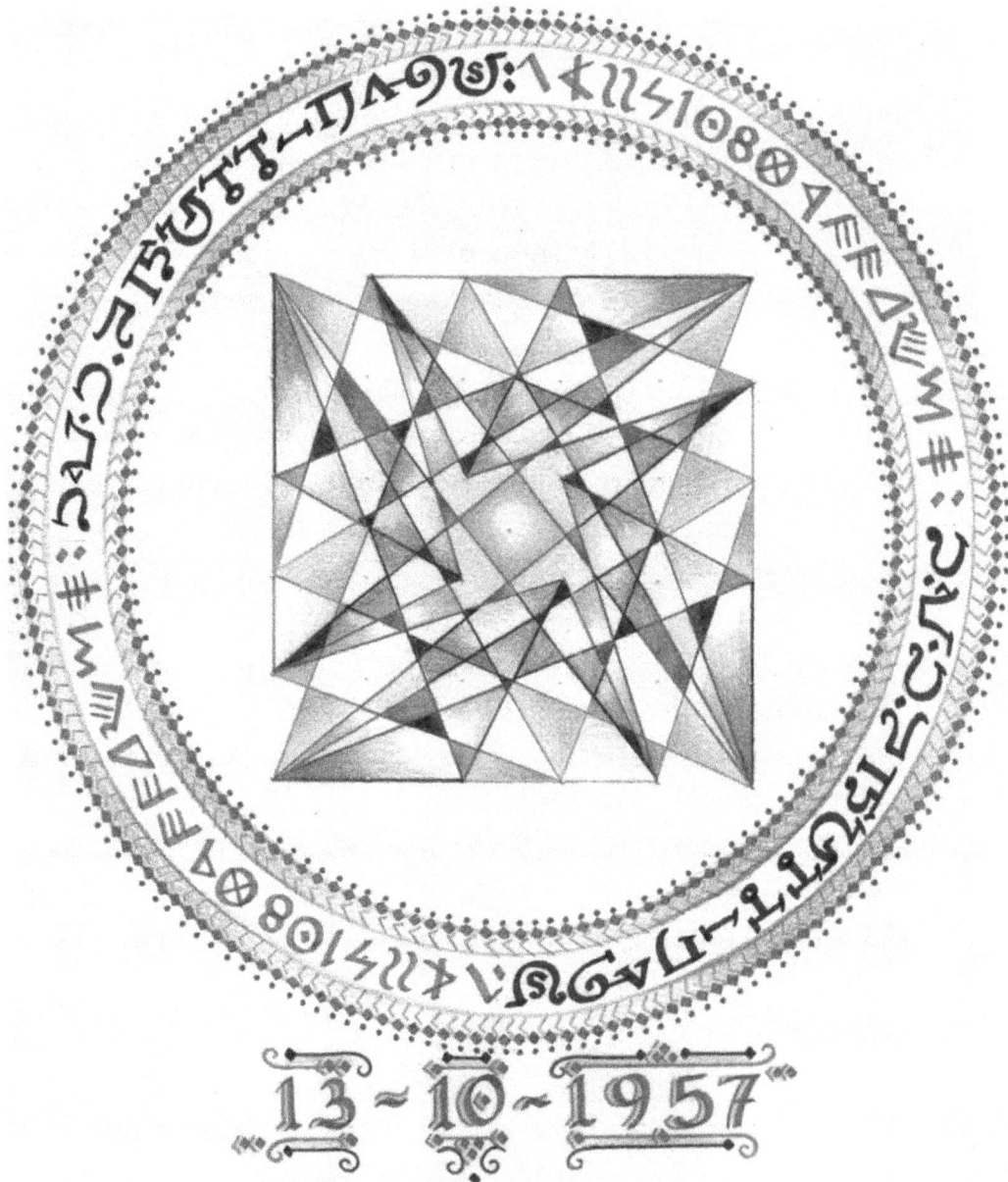

Soul Signature 1,
Yantra by Jain 108 and Art by Lily Moses.

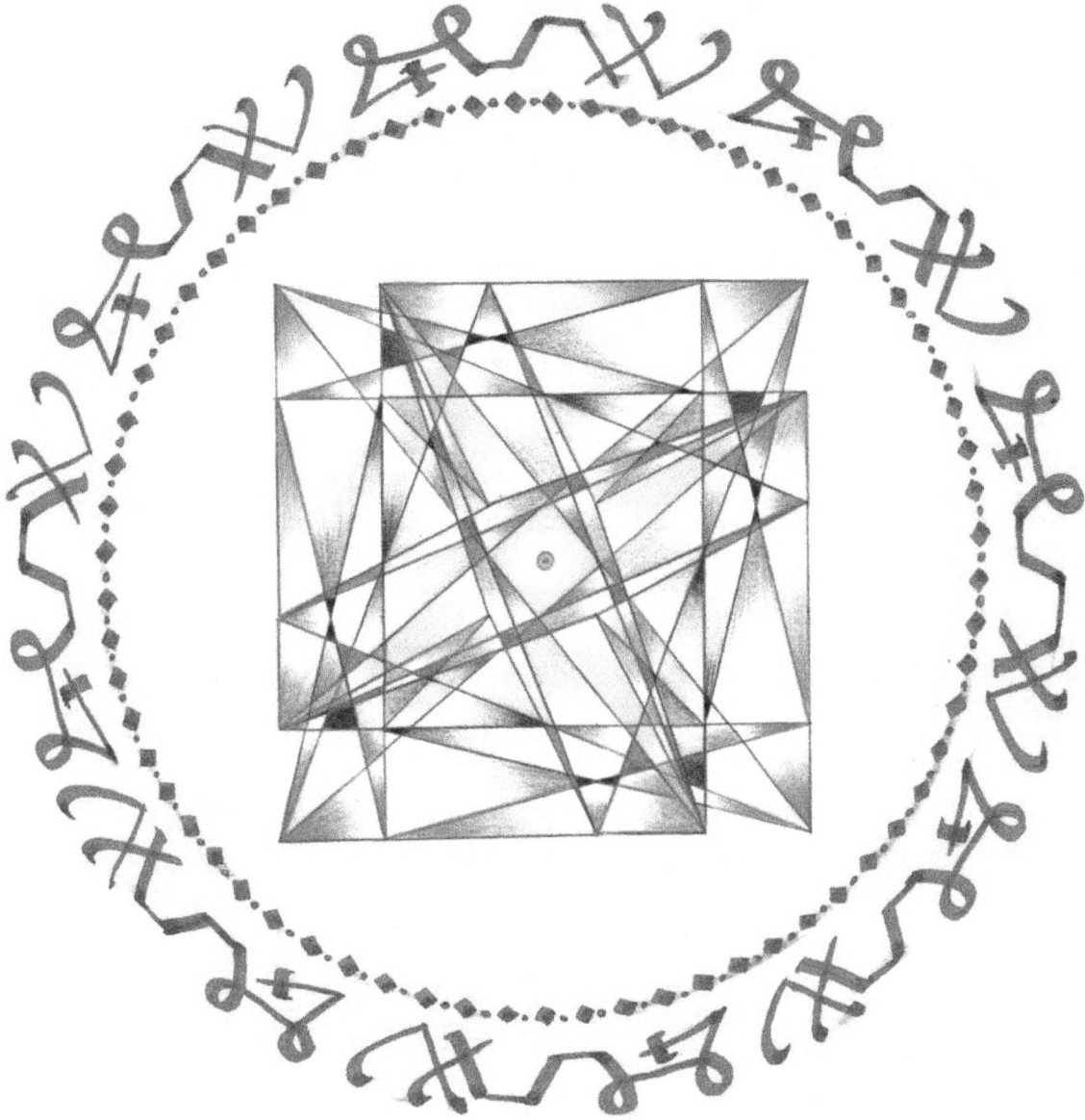

Soul Signature 2, yantra drawn and coloured by Jain 108.

JAIN 108   SOUL SIGNATURE   LILY MOSES

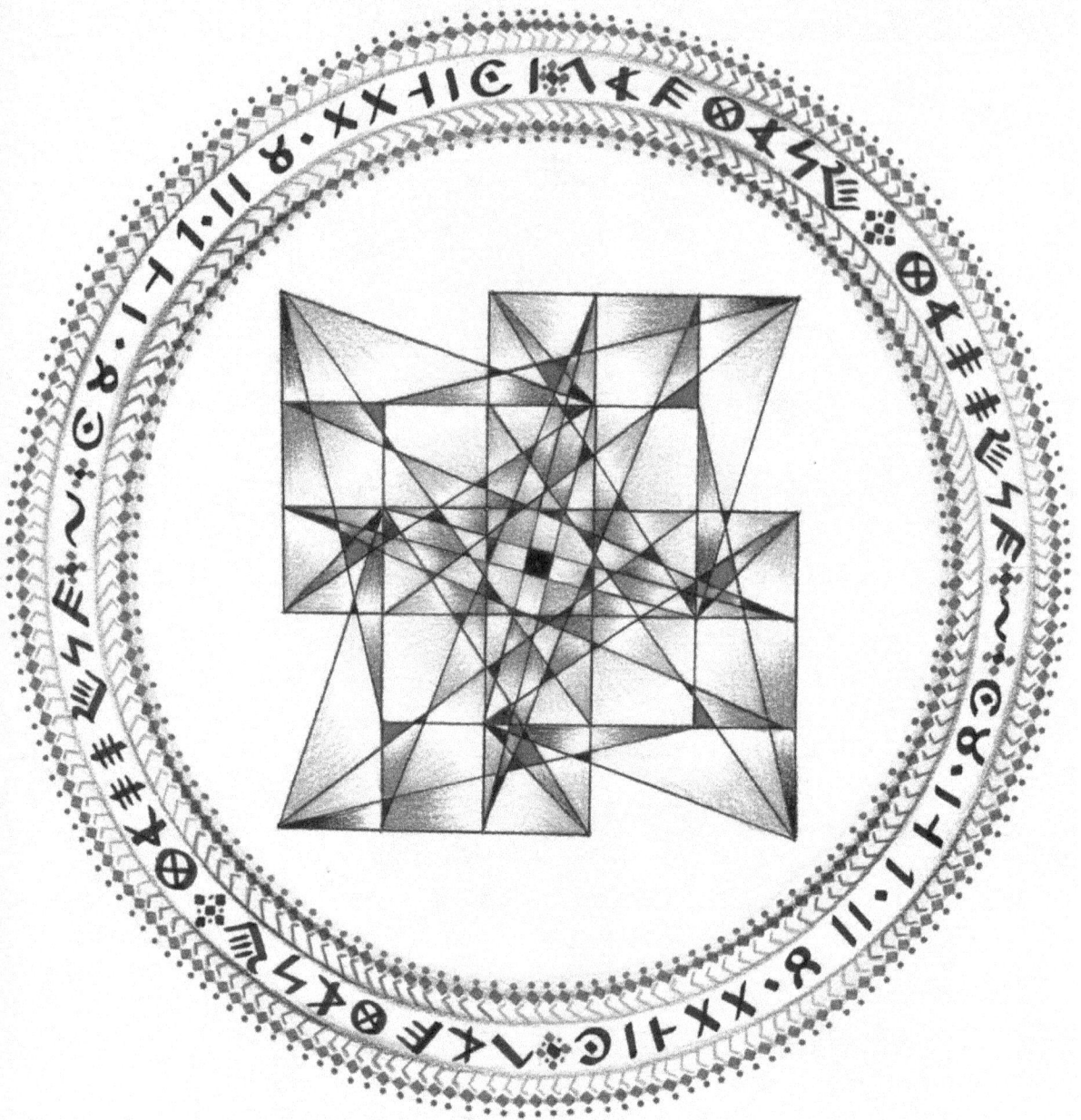

Soul Signature 3,
Yantra by Jain 108 and Art by Lily Moses.

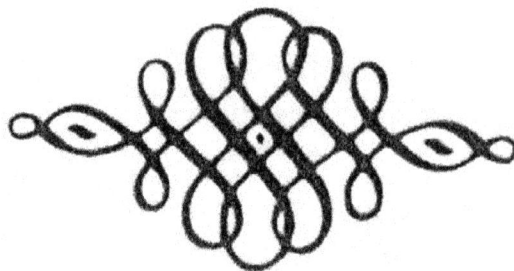

Jain lecturing at a 3 day Utopia Conference, Noosa Heads, Queensland.
Oct 2009. As a guest speaker, I was teaching the audience
of 100 people how to breathe into their Heart,
called the EarthHeart Meditation (aka Eartheart).
Notice the wheel of 24 Phi Code numbers
demonstrating the 12 Pairs of 9 that sum to 108.

Thanks to **Aurelien Floret** for the graphic arts creation of both the front and back cover of this book.
The image above is called "Hypercube" by Aurelien.
View his amazing collection of digitally created designs and fractal art
in full colour, in his "Gallery 108" by viewing his website:
www.luminaya.com

JAIN 108
Mathemagics Φ

$3^3 + 4^4 + 5^5 = 6^x$

1,1,2,3,5,8,4,3,7,1,8,9,8,8,7,6,4,1,5,6,2,8,1,9

Vedic Maths and Sacred Geometry
Day Seminars for Teens, Juniors and Adults

SCHEDULE

10/3/2012
VEDIC MATHEMATICS
Rapid Mental Calculation –
No more calculators!

17/3/2012 (9-12 year olds)
MATHEMAGICS FOR JUNIORS
A fun introduction to magic squares,
magic fingers & number art

24/3/2012
THE DIVINE PROPORTION
The Fibonacci numbers and the
living mathematics of nature

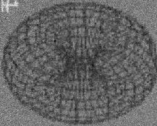

31/3/2012
3-DIMENSIONAL GEOMETRY
The 5 platonic solids – key to
atomic structure

$$\text{Phi } (\phi) = \frac{1 + \sqrt{5}}{2}$$
$$= 1.618033\ldots$$

34,560

$3435 = 3^3 + 4^4 + 3^3 + 5^5$

Jain@JainMathematics.com

108^2? Answer $= 108 + 8/8^2 = 116/64 = 11,664$

(all events held @ the Shearwater Steiner School, Mullumbimby, NSW)

Jain 108's 5 interactive seminar titles
nb: The Cuboctahedron is in the centre of the Pentacle!

The Art of Number

● This book is a distillation of all other 15 books hitherto authored by Jain. It's a thesis that explores the mathematical derivation or origins of sacred symbols like the Golden Spiral (Ram's Horn), the VW symbol, the Pentagram, the Star of David, the Seal of Solomon, Magic Squares and much more.

● Its basic premise is to view Mathematics as Pictures or Power Art, that is recognized as Feminine, Right Brain Mathematics. Its like the reader is putting on X-Ray goggles to see Numbers as Shapes: the Universal Language of Pattern Recognition.

● Jain explores sequences like the Fibonacci Sequence and the Binary Sequence and magically transforms them into exquisite Art.

● Some sequences extracted from the common Multiplication Tables are simplified by Digital Reduction or Compression to single digits and reveal hidden Atomic Art.

● This book is a rare collection of the finest patterns available, and has never before been comprehensively compiled in such a visually stunning way.

● Great for teenage students wanting to further their love of Numbers and great for adults who are willing to learn again.

● Jain believes that if most students who struggled with mathematics were to simply have studied these mathematical patterns, and not learnt their algebra and trigonometry as required, would have seen life through another lens and developed a deeper appreciation for the Beauty of Mathematics.

● "If I were to leave this planet and leave one gift that would best serve me to be remembered by or to share gems of knowledge, it would no doubt be this Compendium. I am excited to release these gems in the exciting year of 2012, a time when many hidden or esoteric studies are being raised into full view. I believe that these simple Mathematical Codes are Celestial Transcripts and are part of our Ascension Process." Jain

Jain 108

jainmathemagics.com

(Back Cover for this book: THE ART OF NUMBER)

249

forthcoming: **SET of 12 DECALS**
(small and large Adhesive Transparencies for the International market,
based on the 12 main da Vinci-like psycho-active Codes in this book: "The Art Of Number").

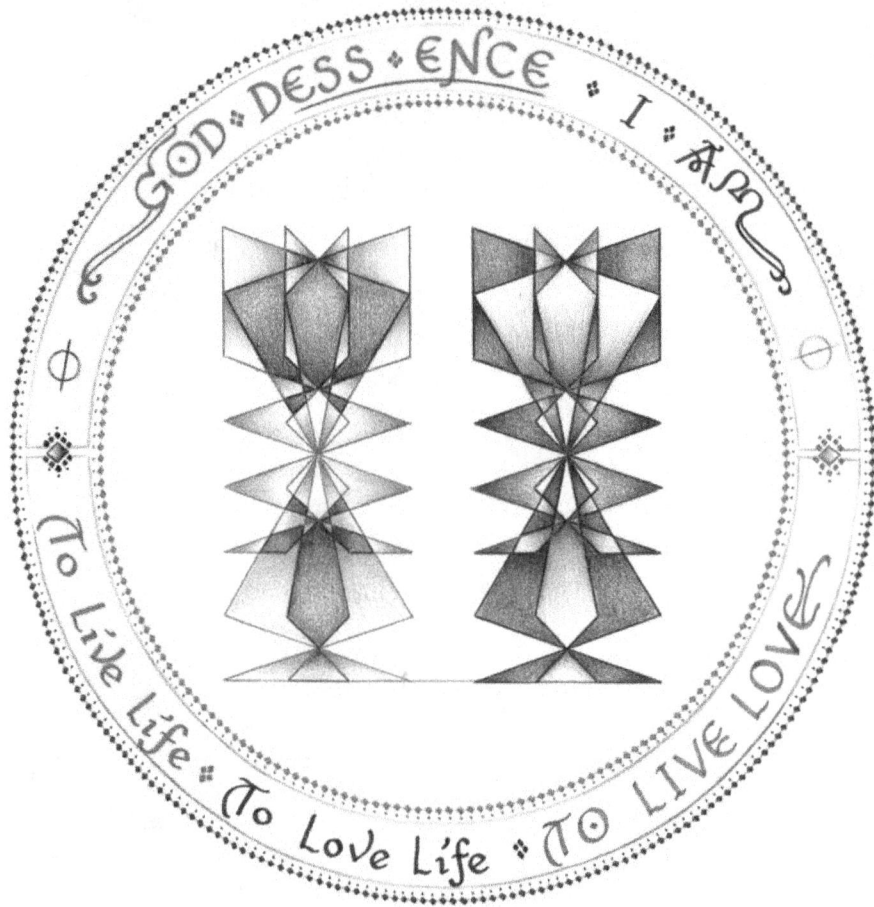

The Benjamin Franklin Magic Square of 8x8 at 0 Degrees.
(Designs above and below are drawn by Jain, but colouring in is by Lily Moses, 2012)

The Benjamin Franklin Magic Square of 8x8 at 4x90 Degrees

52	61	4	13	20	29	36	45
14	3	62	51	46	35	30	19
53	60	5	12	21	28	37	44
11	6	59	54	43	38	27	22
55	58	7	10	23	26	39	42
9	8	57	56	41	40	25	24
50	63	2	15	18	31	34	47
16	1	64	49	48	33	32	17

The Benjamin Franklin Magic Square of 8x8
whose Constant or Magic Sum of all Columns and Rows (not Diagonals) = 260

INDEX

I N D E X + Number Harmonics + Sequences

nb: any page numbers <u>underlined</u> means that this reference is an **image** or a **chart**, not text.

LEGEND:
~ "aka" means "Also Known As".
~ "Ch" means "Chapter".
~ "M. Sq." means "Magic Square".
~ "Nos." means "Numbers".
~ "Seq" means "Sequence".

THIS INDEX CONTAINS 3 PARTS:

- INDEX OF WORDS

- INDEX OF NUMBERS

- INDEX OF DIGITALLY COMPRESSED SEQUENCES

Art by Jain, 1999,

Mural of Hathor,

oil on canvas,

(a 4th Dimensional Venusian, cow-earred being; a sonic and frequency master).

INDEX
OF WORDS

Maltese Cross — (see Prime Number Cross), 152 (Salmon Tails),

Mandala — 208,

Manjushri — 127,

Masonic — 15, 130, 160,

Mastercard — 99,

Mathematics — 124 (Intuitive), 126 (mathematical wounds),

Mathematical Confidence — 20, 55,

Matrix — 8, 61, 169, 183-184, 229, 240-241,

Mattang — 161,

Mayan Calendar — 122, 180 (conference),

McCanney, James — 140,

McGowan, Kathleen — 160 (red Prime Number Cross),

Meditation — 106 (EarthHeart),

Memory — 55 (Power), 103 (Shape Stores Memory),

Memory Power — 20,

Merkabah — 135,

Meru Prastera — 116-125 (aka Pascal's Triangle),

Metatron — 229 (Metaton's Cube), 237, 240-244 (Metatronic Script),

Method Of Transposition — 182 (creates M. Sq. of 3),

Micro-Psi — 74, 173-179 (see Leadbeater),

Mirror Axis — 26-27, 30, 209,

Mitosis — 129, 135,

Modulus System — 13 (Decimal System is really Mod 9),

Mona Lisa — 91,

Moon — 206,

Moses, Lily — 240-244 (Soul Signature), 250 (Magic Square of 8x8),

Mueller, E.H. — 73-74,

Mural by Jain — 75 (platinum crystal),

Multiplication Table — 18-19, 56-86 (Chap 3), 57-60 (4 Times Table), 268-269 (digitally compression of all number sequences from 1 to 9),

N

Nabokov (Spiral) — 22, 40-42, 47,

Name — 240-244 (Soul Signature),

NASA — 140,

Naudin — 102,

Natural Square — 181 (of 3x3), 189,

Nature — 9-11, 45,

Nautilus Shell — 10, 91-96,

Negative Space — 215,

Newton, Issac — 165,

Nine Point Circle — 24-29,

Now — 122-123,

Nuclear Physics — 188,

Numerical Reduction — (see Digital Compression),

O

Octagram — 19,

Octahedron — 70 (rutile crystal), 229 (in Metaton's Cube),

Occult Chemistry — 173-179 (see Leadbeater),

Odd Numbers — 48-55,

Off-Centre — 31,

Oneness — 19,

INDEX
OF
NUMBERS

Of Jain), <u>232-234</u> (12 spheres around 1), <u>238</u> (Kathara Grid: True Tree Of Life),

13 — 122 (Mayan Calendar 13:20), <u>229</u> (Fruit of Life creates Metatron's Cube), <u>233</u> (12 spheres around 1),

16 — 207 (Star of David),

17 — 13 (Jain's 17th Sutra),

20 — <u>98-99</u> (Dodecahedron + Stellation), 111 (sum of some columns in 60 Final Fibonacci Digits), <u>219-221</u> (Indian Magic Triangle),

21 — 9-11, <u>90</u> (sunflower floret 21:34), 104 (Fractal Compression 21:34), 135,

22 — <u>236</u>, (Letters of Alphabet),

24 — <u>15</u>, <u>21</u> (3 Phi Codes in a 3x8 array), 33, <u>37-38</u> (Triangular Vectors), <u>100-106</u>, <u>108-115</u> (12:24:60 code), 124, <u>151</u> (Chart of Prime Distribution), <u>156-161</u> (Primes), 207 (Star of David), 217-<u>218</u>, 158 (IcosaTetraGon), <u>238</u> Stargate), 245 (EarthHeart),

24 Repeating Pattern — 15, 101,

25 — 111 (sum of some columns in 60 Final Fibonacci Digits),

27 — <u>231</u> (tetrahedra),

28 — 222,

30 — 112 (sum of some columns in 60 Final Fibonacci Digits),

32 — <u>93</u> (32 Degrees of main Diagonal in Phi Spiral in Phi Rectangle),

34 — 9, <u>90</u> (sunflower floret 21:34), 104 (Fractal Compression 21:34), 135,

36 — 222,

60 — <u>108-115</u> (12:24:60 code), 115-116, 267 (Phi In Trinitized Vesica Piscis),

64 — See Nassim Haramein, <u>231</u> (tetrahedra + cubic sequence), 234 (tetrahedra), <u>239</u>, (I-Ching),

72 — 218,

81 — <u>66-67</u> (Grid of 81 Dots of Times Table),

90° — <u>173-179</u> (M. Sq.3 alternate rotations), 200, <u>213-214</u> (Coptic Cross),

108 — 9-21 (108 Codes), 13 (108 expressed in Base 2 Binary), 15, <u>21</u> (3 Phi Codes in a 3x8 array), 67, 84, 93, 102, 112, 245 (EarthHeart),

111 — 19, <u>240-244</u> (Soul Signature),

125 — 231 (part of cubic sequence),

144 — 15-16,

147 — Solfeggio Scales: see "3 Phi Codes", — <u>21</u>,

216 — 231 (part of cubic sequence),

258 — Solfeggio Scales: see "3 Phi Codes", <u>21</u>,

260 — 250-251 (forthcoming Decal designs by Jain of Ben Franklin),

280 — 111 (sum of some columns in 60 Final Fibonacci Digits),

300 — <u>113-114</u> (Periodicity of Final Two Fibonacci Digits), 115-116,

343 — 231 (cubic sequence),

366 — 123 (days in a sidereal year),

369 — Solfeggio Scales: see "3 Phi Codes", <u>21</u>, 28 (Missing Gap Sequence in Binary Code in 9 Point Circle),193 (Tesla 3-6-9 Sequence),

512 — 11, 129 (cells forming original Torus), 231 (part of cubic sequence),

666 — 19, <u>165</u> (Queen's Crown),

729 — 231 (part of cubic sequence),

999 — 28 (pattern in Binary Code),

1,000 — 129(spirals based on equal spin),

1,008 — 129 (spirals based not on equal spin),

1,500 — 113-114 (Periodicity for the Final 3 Digits of fib seq), 115-116,

2012 — 122 (2012 Prophesies),

15,000 — 113-114 (Periodicity for the Final 4 Digits of fib seq), 115-116,

150,000 — 113-114 (Periodicity for the Final 5 Digits of fib seq), 115-116,

786,432 — 16 (Hertz of electrified quartz crystal),

DID YOU KNOW	Phi In Trinitized Vesica Piscis
	That the 2 circled Vesica Piscis, where 2 circles go through one another's centres, the supposed "Mother Of All Form" (√2, √3 and √5) does not contain the Phi Ratio. It does though contain the 3 important Square Root Harmonics of Root 2, Root 3 and Root 5, as shown on the left. To my surprise, I recently discovered in a beautiful book where the True Vesica Pisces reveals directly the Phi Ratio:
	Here it is. To get **Phi**, we had to divide the diameter of one circle into 3, so that 3 smaller circles can fit into the larger one. This <u>Trinitization</u> of Consciousness was the Key, just like **The 3 Phi Codes** (see page 21) referenced in this book and forthcoming ones. This is genius. CB:AC :: 1:1.618... The overall Grid to contain this geometry is based on 6x10 = 60 cells. (image sourced from "**The Glorious Phi**" by A. S. Posamentier and I. Lehmann, but the actual geometry was popularized by Hans Walser. Thank You so much, I had waited for this diagram for about 30 years. Use it in your Art or Designs.

INDEX
OF DIGITALLY COMPRESSED
SEQUENCES

("P" = Periodicity = How many numbers in that Sequence)

1, 1, 2, 3, 5, 8, 4, 3, 7, 1, 8, 9, 8, 8, 7, 6, 4, 1, 5, 6, 2, 8, 1, 9
Phi Code 1: (1,1,2)
— The Compression of the primal Fibonacci Sequence;
The 1st of the 3 possible Dials. P=24

1, 2, 3, 4, 5, 6, 7, 8, 9
Digitally Compressed 1x Times Table. P=9

1, 2, 4, 8, 7, 5
Binary Code. P=6

1, 2, 6, 3, 6, 6, 9
Digitally Compressed Sequence of Phibonacci Products, from a previous book on phi. P=7

1, 3, 4, 7, 2, 9, 2, 2, 4, 6, 1, 7, 8, 6, 5, 2, 7, 9, 7, 7, 5, 3, 8, 2
Phi Code 2 PC2: (1,3,4)
— The Powers Of Phi 108) The Compression of another Fibonacci Sequence.
The 2nd of the 3 possible Dials. P=24

1, 4, 5, 9, 5, 5, 1, 6, 7, 4, 2, 6, 8, 5, 4, 9, 4, 4, 8, 3, 2, 5, 7, 3
— Phi Code 3: PC3: (1,4,5)
The Compression of another Fibonacci Sequence;
The 3rd of the 3 possible Dials. P=24

1, 4, 9, 7, 7, 9, 4, 1, 9
from the n^2 = (Squared Numbers Sequence):
1, 4, 9, 16, 25, 36, 49, 64, 81, 100, 121, 144... P=9

1, 8, 9
from the n^3 = (Cubic Numbers Sequence):
1, 8, 27, 64, 125, 216, 343, 512, 729, 1000... P=3

2, 4, 6, 8, 1, 3, 5, 7, 9
Digitally Compressed 2x Times Table. P=9

3, 6, 9
Digitally Compressed 3x Times Table. P=3

4, 8, 3, 7, 2, 6, 1, 5, 9
Digitally Compressed 4 Times Table). P=9

5, 1, 6, 2, 7, 3, 8, 4, 9
Digitally Compressed 5x Times Table. P=9

6, 3, 9
Digitally Compressed 6x Times Table. P=3

7, 5, 3, 1, 8, 6, 4, 2, 9
Digitally Compressed 7x Times Table. P=9

8, 7, 6, 5, 4, 3, 2, 1, 9
Digitally Compressed 8x Times Table. P=9

9, 9, 9
The Binary Code cut in half and Pairs adding to 9. P=3

12:24:60
Time Code of the Fibonacci Sequence